Quentin Hiernaux, Corentin Tresnie

Andrea Cesalpino's *De Plantis Libri XVI* (1583) and the Transformation of Medical Botany in the 16th Century

Medical Traditions

Edited by
Alain Touwaide

Volume 9

Quentin Hiernaux, Corentin Tresnie

Andrea Cesalpino's *De Plantis Libri XVI* (1583) and the Transformation of Medical Botany in the 16th Century

Edition, Translation, and Commentary on Book I

DE GRUYTER

ISBN 978-3-11-100016-9
e-ISBN (PDF) 978-3-11-100110-4
ISSN 2567-6938

Library of Congress Control Number: 2023937889

Bibliographic information published by the Deutsche Nationalbibliothek
The Deutsche Nationalbibliothek lists this publication in the Deutsche Nationalbibliografie; detailed bibliographic data are available in the Internet at http://dnb.dnb.de.

Cover image: Collage of illustrations in medical treatises from the 1st to the 16th century CE, from Greece and Rome to the Arabic World and China (1. peaches from a Japanese edition from the Li Shi Zhen, Bencao Gangmu, A.D. 1714; 2. plant from the manuscript of Padova, Biblioteca Seminario, 194, Constantinople, ca. A.D. 1430; 3. scene of uroscopy from the manuscript of Paris, Bibliotheque nationale de France, grec, 2294, 15th century; 4. a man climbing on a tree from a manuscript of Dioscorides, Arabic, Bologna, Biblioteca universitaria, 2954, A.D. 1244)
Printing and Binding: LSC Communications, United States

www.degruyter.com

DE PLANTIS
LIBRI XVI.
ANDREAE CAESALPINI
ARETINI,
Medici clariſsimi, doctiſsimiq; atque
Philoſophi celeberrimi, ac
ſubtiliſsimi.

AD SERENISSIMUM FRANCISCUM
Medicem, Magnum Aetruriae Ducem.

FLORENTIAE,
Apud Georgium Mareſcottum.
MDLXXXIII.

Figure 1: Front cover of Andrea Cesalpino's *De plantis libri XVI*, Georgius Marescottus, 1583.

Foreword

Andrea Cesalpino's *De Plantis* (*On Plants*) (1583) was a landmark in biological systematics. In its pages, Cesalpino was the first to provide detailed criteria for determining botanical species. Because plants vary in how they grow and develop, based on factors of time, place, and circumstance, he focused his definition of plant species on their reproductive organs: primarily, the fruit, with the flower providing a secondary form of differentiation. While he and his contemporaries were unaware of plant sexuality, the plant groups he established on the basis of his morphological studies mark an important step beyond the folkbiological groupings found in his predecessors' works. Recognized by successors such Joachim Jung, John Ray, and Carl Linnaeus as a key innovation, Cesalpino's work has grown increasingly inaccessible; it was never translated into a modern language, and even Latin readers can be misled by its technical terminology. With their edition, translation, and commentary of the first book of *On Plants*, Quentin Hiernaux and Corentin Tresnie have provided a significant contribution to the history of early modern science and the development of biology.

Cesalpino's work united Aristotelian natural philosophy with the empiricism of 16th-century natural history. As professor at the University of Pisa, Cesalpino was intimately familiar with Aristotle's works, which were the basis of philosophical education. His interests included writing on *Peripatetic Questions*, the title of a book he published in 1571. While other Renaissance naturalists were also steeped in Aristotelian philosophy, Cesalpino gave it more thought than many, and the title page of *On Plants* proclaimed that he was not only "a most famous and learned Physician" but also "a most celebrated and subtle Philosopher". At the same time, he possessed broad practical knowledge of plants: he had studied with the famous teacher of medical botany Luca Ghini, whom he succeeded as director of the botanical garden in Pisa. More than any other botanist of the 16th century, he strove to emulate Theophrastus, Aristotle's associate and successor at the Lyceum, in taking a philosophical approach to empirical botany.

As his predecessors had done, Cesalpino recognized that some aspects of plant morphology were highly variable. Therefore, he sought a way to differentiate, in Aristotelian terms, "accidental" differences between individuals from the universal characters that defined the essence of a species. In *On the Soul*, a work concerned with the natures of living creatures, Aristotle argued that plants possessed souls with the nutritive functions of growth and reproduction, but they lacked the further powers of sensation and motion found in animals (as well as reason, unique to humans). Within this framework, Cesalpino posited that the nutritive functions of plants' souls, particularly reproduction, provided the key to universal characteristics, since reproduction was how each species perpetuated itself. Hence his focus on the reproductive organs, and particularly the fruit or seed. Cesalpino's concept of species took Aristotle's basic principle that "like produces like" and refined it, conceptually and empirically, to produce the first detailed biological concept of a species and to lay the foundations for systematic groupings of

https://doi.org/10.1515/9783111001104-201

species in larger taxa. With the resources of a botanical garden and herbaria, Cesalpino was able to examine a broad range of plants' fruit and flowers to establish 32 distinctive groups of plants, united by reproductive similarity despite many other significant differences.

A distinctive aspect of Cesalpino's approach is that despite his medical background, he included the medicinal virtues of plants among their accidental qualities, not the substantial qualities that define a species. Renaissance herbals, following the model of Dioscorides, emphasized the healing virtues of plants and their uses as ingredients in compound medicines, sometimes using those virtues as a basis for grouping plants together. In *De plantis*, Cesalpino rejected that approach: he acknowledged that plants that belonged to a common grouping often possessed similar medicinal virtues, but the latter were a consequence of the plants' relationship, not the grounds for establishing one. Seen in retrospect, Cesalpino's approach appears to mark the beginnings of an "emancipation" of the science of botany from medicinal herbalism. In his dedication letter serving as a preface of the work to Francesco de' Medici, Grand Duke of Tuscany (figure 2), Cesalpino emphasized the variety and beauty of plants as sufficient reason to study them, aside from their practical uses. Yet his frequent, though not systematic, references to medicinal virtues, and their inclusion in the index to *De plantis*, suggests that he saw his task as one of reform.

Cesalpino was no iconoclast. Like other participants in the "scientific Renaissance" of the 15th and 16th centuries, he set out to clarify and refine knowledge that had been imperfectly transmitted from classical antiquity, while expanding its empirical scope. But neither was he content to merely follow in the footsteps of the Ancients. He aimed to emulate them, not imitate them, and thereby achieve something new. As with Nicolaus Copernicus and Andreas Vesalius, the result was a substantial contribution to knowledge of nature. His ideas were very much products of their time, but he was the ideal person to generate them.

Cesalpino's importance has long been recognized, but as reading knowledge of Latin has become rare, fewer historians and biologists have been able to directly access his writings. Furthermore, botanical terminology has changed since Cesalpino's day: for example, he used the word *stamen* to designate both the style and the filament. Hence it is timely that this new edition of the first book of *De plantis*, with translation and commentary, has been published. In the first book, Cesalpino sets out his theoretical and general views about plants. Readers with no Latin will learn a wealth about Cesalpino's contributions to botanical history from the clear translation and insightful commentary. Those who can read Latin but are unfamiliar with the technical terminology of the period, including terms that Cesalpino used in novel ways, will also benefit from close comparison of the text and translation. A helpful index of the Latin plant names in Book I, including references in other books, the Latin binomial name (when this can be identified) and common English name (when it exists), will guide the reader wishing to compare Cesalpino's botany with modern specimens. The present publication does a

great service to botanists, historians of botany, and intellectual historians who can now engage with Cesalpino's ideas accessibly and with expert guidance.

Brian W. Ogilvie
Professor of History
University of Massachusetts Amherst

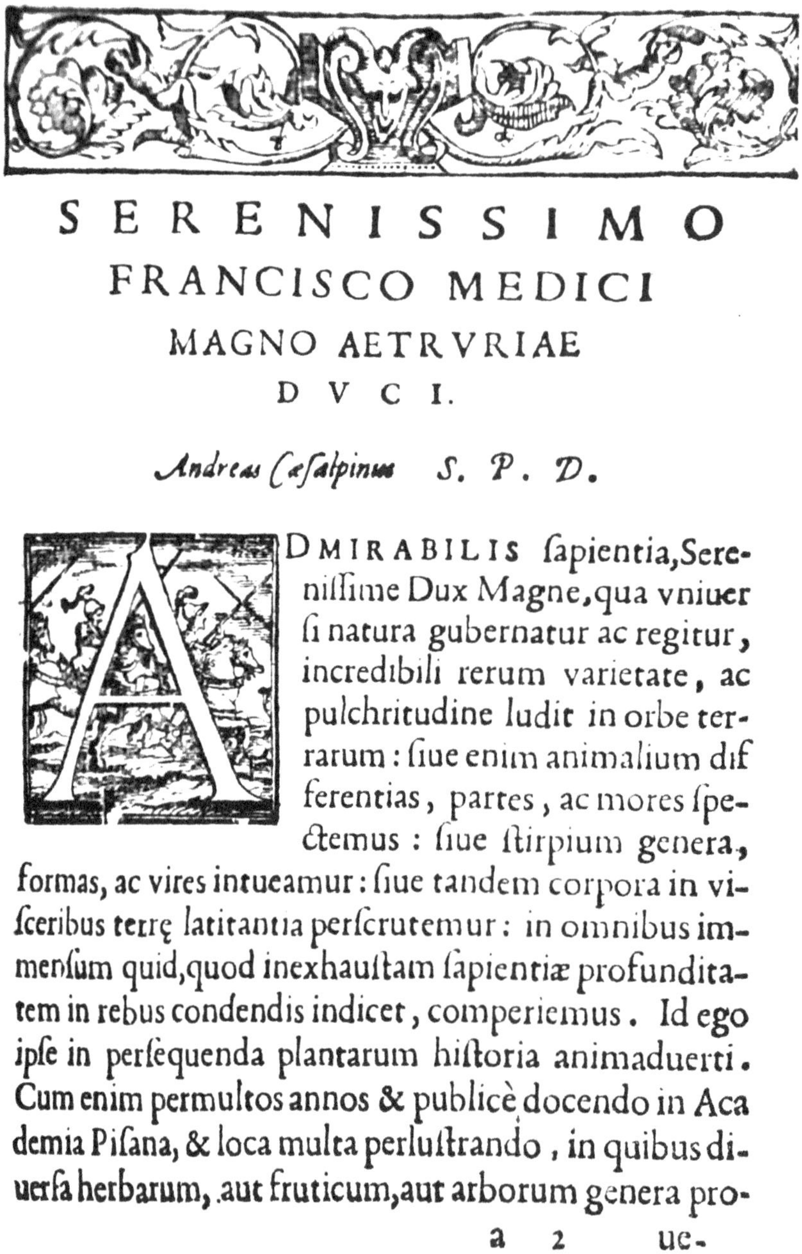

SERENISSIMO
FRANCISCO MEDICI
MAGNO AETRVRIAE
DVCI.

Andreas Cæſalpinus S. P. D.

ADMIRABILIS ſapientia, Sereniſſime Dux Magne, qua vniuerſi natura gubernatur ac regitur, incredibili rerum varietate, ac pulchritudine ludit in orbe terrarum: ſiue enim animalium differentias, partes, ac mores ſpectemus: ſiue ſtirpium genera, formas, ac vires intueamur: ſiue tandem corpora in viſceribus terrę latitantia perſcrutemur: in omnibus immenſum quid, quod inexhauſtam ſapientiæ profunditatem in rebus condendis indicet, comperiemus. Id ego ipſe in perſequenda plantarum hiſtoria animaduerti. Cum enim permultos annos & publicè docendo in Academia Piſana, & loca multa perluſtrando, in quibus diuerſa herbarum, aut fruticum, aut arborum genera pro-

a 2 ue-

Figure 2: First page of the dedication letter of *De plantis* to the Grand Duke of Tuscany, Francisco de' Medici, 1583.

Table of Contents

Foreword —— **VII**
Acknowledgments —— **XIII**
List of Figures —— **XV**

Introduction —— **1**

Part I: An Introductory Essay
Biography of Andrea Cesalpino —— **5**
Historical Context of *De plantis* —— **7**
Outline of Book I —— **18**
Cesalpino's Method, Intellectual Innovations, and Influences —— **20**
Finalism, Nature, and God —— **26**
The Analogy —— **32**
Positioning of Cesalpino in the History of Classification —— **36**
Status of the Illustration in *De plantis* —— **47**
The Dried Herbarium of Cesalpino (*Hortus siccus*) —— **49**
The Physiology of Cesalpino —— **51**

Part II: Text and Translation
Notes on Editing —— **55**
Notes on the Translation —— **56**
Table of Chapters —— **57**
Latin Text and Facing English Translation of Book I
Chapter 1 [The properties and parts of plants: §1–14] —— **58/59**
Chapter 2 [Nutrition: §15–26] —— **64/65**
Chapter 3 [Development and growth: embryo, bud, shoot and bark: §27–37] —— **72/73**
Chapter 4 [Growth and development: heart, stem, soboles and leaves: §38–47] —— **80/81**
Chapter 5 [Vegetative reproduction: §48–54] —— **86/87**
Chapter 6 [Seeds: §55–63] —— **90/91**
Chapter 7 [Flowers: §64–76] —— **98/99**
Chapter 8 [Fruits and seed coats: §77–85] —— **110/111**
Chapter 9 [The pericarp: §86–92] —— **118/119**
Chapter 10 [The parts dedicated to the protection of the fruit and fructification: §93–113] —— **124/125**
Chapter 11 [The secondary parts: §114–117] —— **138/139**
Chapter 12 [The four main genera and their divisions: §118–134] —— **142/143**
Chapter 13 [Criteria for defining subgenera and species: §135–144] —— **152/153**
Chapter 14 [Subdivisions based on the reproductive operation: §145–157] —— **158/159**

Part III: Commentary

Introduction to the Commentary — **169**
Commentary of the Fourteen Chapters of Book 1 — **170**
Chapter 1. The properties and parts of plants: §1–14 — **170**
Chapter 2. Nutrition: §15–26 — **174**
Chapter 3. Development and growth: embryo, bud, shoot and bark: §27–37 — **177**
Chapter 4. Growth and development: heart, stem, soboles and leaves: §38–47 — **181**
Chapter 5. Vegetative reproduction: §48–54 — **184**
Chapter 6. Seeds: §55–63 — **186**
Chapter 7. Flowers: §64–76 — **190**
Chapter 8. Fruits and seed coats: §77–85 — **199**
Chapter 9. The pericarp: §86–92 — **202**
Chapter 10. The parts dedicated to the protection of the fruit and fructification: §93–113 — **203**
Chapter 11. The secondary parts: §114–117 — **209**
Chapter 12. The four main genera and their divisions: §118–134 — **216**
Chapter 13. Criteria for defining subgenera and species: §135–144 — **220**
Chapter 14. Subdivisions based on the reproductive operation: §145–157 — **222**

Part IV: Appendices

Bibliography — **227**
Index of Plant Names in *De plantis*, Book I — **239**
Thematic Index — **227**

Acknowledgments

This work is the result of years of research conducted at the Philosophy research center (PHI) of the Université libre de Bruxelles (ULB), made possible by the National Fund for Scientific Research of Belgium (FRS-FNRS). We thank the FNRS and the ULB for their financial support.

We presented parts of the topic and preliminary results in several conferences, and our thanks go to the organizers of these events, especially to Dr Fabrizio Baldassarri, who coordinated a panel entitled "Andrea Cesalpino: A Physician, Philosopher, and Botanist in Renaissance Italy" at the 2021 Renaissance Society of America (RSA) Conference, and the workshop "Andrea Cesalpino: An Aristotelian Natural Philosopher in the Renaissance" in 2022 with Prof. Craig Martin. We also thank Prof. Sarah Carvallo and Prof. Arnaud Macé for inviting us to their conference "Analogies et modèles végétaux en médecine dans l'Antiquité et l'Âge classique" in 2019.

We warmly thank the Meise Botanic Garden, its library, and especially historian of botany Dr Denis Diagre for his support and for his suggestions on the draft of this manuscript. We thank the Cambridge University Library and its rare books service for allowing us to consult their two exemplars of Cesalpino's *De plantis libri XVI*.

We acknowledge Nicolas Hiernaux for his talented infographic adaptation of our drawings published in this book.

Special thanks go to the editor of the series Medical Traditions, Prof. Alain Touwaide, not only for this invitation to submit the volume for publication in the series, but also, and above all, for his constant guidance during the preparation of the manuscript. It went well beyond editing, and allowed us to refine our understanding of Cesalpino's work, improve the present work, and deepen our research. All remaining errors and limitations are ours.

Quentin Hiernaux, Research Associate at the Scientific Fund for Research (Université libre de Bruxelles), Scientific collaborator at Meise Botanic Garden,

Corentin Tresnie, Research Fellow at the Scientific Fund for Research (Université libre de Bruxelles and KU Leuven)

https://doi.org/10.1515/9783111001104-202

List of figures

Figure 1: Front cover of Andrea Cesalpino's *De plantis libri XVI*, Georgius Marescottus, 1583. Credits: Cambridge University Library rare books collection, public domain.

Figure 2: First page of the dedication letter of *De plantis* to the Grand Duke of Tuscany, Francisco de' Medici, 1583. Credits: Cambridge University Library rare books collection, public domain.

Figure 3: Portrait of Andrea Cesalpino by G. Zocchi, engraved by G. Allegrini, 1765. Credits: Collection of Meise botanic garden library, public domain.

Figure 4: Plan of the botanic garden of Pisa. Credits: plates designed by Tilli, and together with plans, engraved by Mogalli, 1723, public domain.

Figure 5: Bremekamp 1952 table about Cesalpino's taxonomic groups 1/2. Credits: John Wiley and Sons, reproduced from the original article with permission.

Figure 6: Bremekamp 1952 table about Cesalpino's taxonomic groups 2/2. Credits: John Wiley and Sons, reproduced from the original article with permission.

Figure 7: Bremekamp 1952 diagram linking Cesalpino's taxonomic groups with plant families. Credits: John Wiley and Sons, reproduced from the original article with permission.

Figure 8: Original C drop capital engraved in the first page of the first book of *De plantis libri XVI*, 1583. Credits: Cambridge University Library rare books collection, public domain.

Figure 9: Last page of *De plantis libri XVI*, 1583. Credits: Cambridge University Library rare books collection, public domain.

Figure 10: Illustration of Cesalpino's tree anatomy. Credits: Quentin and Nicolas Hiernaux, 2022.

Figure 11: Current illustration of tree anatomy. Credits: Quentin Hiernaux, 2022.

Figure 12: Illustration of filtration by capillarity. Credits: Quentin and Nicolas Hiernaux, 2022.

Figure 13: Plate of White horehound (*Marrubium vulgare* L.) showing a decussate opposite phyllotaxy. Credits: Franz Eugen Köhler, Köhler's *Medizinal-Pflanzen*, 1897, public domain.

Figure 14: Illustration of the three main phyllotaxies. Credits: Quentin and Nicolas Hiernaux, 2022.

Figure 15: Picture of Rocambole garlic (*Allium scorodoprasum* L.) with cloves germinating at the top of the stem. Credits: Liliane Roubaudi, 2014, PictoFlora/Tela Botanica, Creative Commons BY-SA (2.0).

Figure 16: Plate of Rocambole garlic (*Allium scorodoprasum* L.). Credits: A. Masclef, *Atlas des plantes de France*, 1891, public domain.

Figure 17: Picture of Coral root (*Cardamine bulbifera* (L.) Crantz) showing bulblets along the stem. Credits: Gerard Goujon, 2020, PictoFlora/Tela Botanica, Creative Commons BY-SA (2.0).

Figure 18: Comparison between a monocotyledon and a dicotyledon germination. Credits: Quentin and Nicolas Hiernaux 2022.

Figure 19: Types of flower and ovary implantations. Credits: Quentin and Nicolas Hiernaux, 2022 derived from Ulf Mehlig, Creative Commons BY-SA 2.5.

Figure 20: Drawing of polypetalous (I) and gamopetalous (II) corolla. Credits: Quentin Hiernaux, 2022.

Figure 21: Diagram of floral rings. Credits: Quentin Hiernaux, 2022.

Figure 22: Diagram of a flower. Credits: Mariana Ruiz LadyofHats, 2007. Public domain.

Figure 23: Picture of a Green Hellebore (*Helleborus viridis* L.). Credits: Pablo Behague, 2022, PictoFlora/Tela Botanica, Creative Commons BY-SA (2.0).

Figure 24: Picture of ornithogalum (*Ornithogalum umbellatum* L.). Credits: Mathieu Sinet, 2021, Tela Botanica/PictoFlora, Creative Commons BY-SA (2.0).

Figure 25: Illustration of a sooty mushroom at the end of a wick candle. Credits: Quentin Hiernaux, 2022.

Figure 26: Plate of Meadow-rue (*Thalictrum aquilegiifolium* L.). Credits: Johann Georg Sturm (Painter: Jacob Sturm), *Deutschlands Flora in Abbildungen*, 1796. Public domain.

Figure 27: Picture of a Goat's-head (*Tribulus terrestris* L.)' seed. Credits: Liliane Roubaudi, 2022, PictoFlora/Tela Botanica, Creative Commons BY-SA (2.0).

https://doi.org/10.1515/9783111001104-203

Figure 28: Section of a grape showing the seed and its funiculus. Credits: Liliane Roubaudi, 2000, PictoFlora/Tela Botanica, Creative Commons BY-SA (2.0).
Figure 29: Diagram of a wheat spikelet. Credits: Quentin and Nicolas Hiernaux, 2022.
Figure 30: Plate of Aconitum (*Aconitum Napellus*). Credits: Franz Eugen Köhler, *Köhler's Medizinal-Pflanzen*, 1897. Public domain.
Figure 31: Picture of a ripe rose fruit/seed capsule section showing the down. Credits: Jean-Jacques Houdré, 2015, PictoFlora/Tela Botanica, Creative Commons BY-SA (2.0).
Figure 32: Picture of a rose seed capsule section showing the down. Credits: Thierry Pernot, PictoFlora/Tela Botanica, Creative Commons BY-SA (2.0).
Figure 33: Plate of common corncockle (*Agrostemma gitagho*). Credits: Amédée Masclef, *Atlas des plantes de France*, 1891, public domain.
Figure 34: Plate of greater celandine (*Chelidonium majus* L.). Credits: Franz Eugen Köhler, *Medizinal-Pflanzen*, 1891, public domain.
Figure 35: Left: Pod; Right: Silique (current terminology). Credits: Quentin and Nicolas Hiernaux, 2022.
Figure 36: Diagram of Cesalpino's physiology theory of interlocking layers. Credits: Quentin and Nicolas Hiernaux, 2022.
Figure 37: Diagram of the anatomy of a peach. Credits: LadyofHats 2007, public domain.
Figure 38: Picture of mandrake fruits (*Mandragora officinarum* L.). Credits: Thierry Pernot, 2022, Tela Botanica/PictoFlora, Creative Commons BY-SA (2.0).
Figure 39: Plate of pea (*Pisum sativum* L.). Credits: Otto Wilhelm Thomé, *Flora von Deutschland, Österreich und der Schweiz*, 1885, public domain.
Figure 40: Plate of vetch (*Vicia cracca*). Credits: Johann Georg Sturm (Painter: Jacob Sturm), *Deutschlands Flora in Abbildungen*, 1796, public domain.
Figure 41: Picture of grapevine (*Vitis vinifera* L.) showing tendrils opposed to leaves. Credits: Pierre Bonnet, 2018, PictoFlora/Tela Botanica, Creative Commons BY-SA (2.0).
Figure 42: Picture of clematis (*Clematis flammula* L.) showing petioles winding. Credits: Quentin Hiernaux, 2022.
Figure 43: Picture of ivy (*Hedera helix* L.) showing its claws. Credits: Quentin Hiernaux, 2022.
Figure 44: Picture of silkvine (*Periploca graeca* L.) stem around a tree. Credits: Jean-Claude Bouzat, 2018, PictoFlora/Tela botanica, Creative Commons BY-SA (2.0).
Figure 45: Picture of bindweed (*Convolvulus arvensis*) stem. Credits: Quentin Hiernaux, 2022.
Figure 46: Picture of old man's beards (*Usnea barbata*). Credits: Claudine and Pierre Guezennec, 2012, PictoFlora/Tela Botanica, Creative Commons BY-SA (2.0).
Figure 47: Picture of a *Quercus coccifera* L. gall. Credits: Liliane Roubaudi, 1992, PictoFlora/Tela Botanica, Creative Commons BY-SA (2.0).
Figure 48: Picture of the dog rose gall or Bedeguar. Credits: Stéphanie Georget, 2015, PictoFlora/Tela Botanica, Creative Commons BY-SA (2.0).
Figure 49: Sporangia under a leaf of *Asplenium scolopendrium* L. Credits: Sylvain Piry, 2021, PictoFlora/Tela Botanica, Creative Commons BY-SA (2.0).

Information about the license Creative Commons BY-SA used for several figures are available on: https://creativecommons.org/licenses/by-sa/2.0/fr/.

Table 1: Table of content of *De plantis* Book I.
Table 2: Table of the four main genera.

Introduction

In 1583, Italian botanist, physician, and philosopher Andrea Cesalpino (1524–1603) published *De plantis libri XVI*[1] in sixteen books (*libri*), one of the first early modern treatises to study botany[2] independently from medicine. Cesalpino's approach broke with a long tradition, inherited by Western science from Antiquity, and perpetuated during the Middle Ages and early Renaissance.[3] *De plantis* lays the foundations for scientific systematics through its innovative focus on plant morphology and natural similarities, and has become a milestone in the history of Western botany. It is a precious testimony to the transformation of botanical and physiological knowledge throughout the Middle Ages and into the Renaissance, and reflects the state of Aristotelian philosophy in 16th-century knowledge.[4] It is also the first Renaissance philosophical treatise on plants.[5] Despite its status and considerable influence on later botanists such as Gaspard Bauhin (1560–1624), Joachim Jung (1587–1657) Robert Morison (1620–1683), John Ray (1627–1705), Augustus Quirinus Rivinus (1652–1723), Joseph Pitton de Tournefort (1656–1708), and even Carl Linnaeus (1707–1778) himself,[6] studies based on a reading of the original Latin text remain rare. More importantly, no translation of *De plantis* into a modern language has been published to date. The present work is an edition, translation into English, and analysis of Book I and its context. Book I is particularly significant insofar as it offers a theoretical presentation of Cesalpino's new botany, while the subsequent fifteen books detail the characteristics of about 1500 plants.[7]

This book is divided into four parts. The first is an introductory essay in which we reconstruct Cesalpino's itinerary from Aristotelianism and the ancient and medieval tradition to the Renaissance and the author's own world. The second part is an edition of Cesalpino's Latin text based on the 1583 publication.[8] The Latin text is faced by an English translation in which we have tried to remain as faithful as possible to the orig-

1 Hereafter abbreviated to *De plantis*.

2 From a terminological point of view, the word "botany" was first used towards the end of the 17th century and only became common in the 18th century (Bellorini 2016: 53–54). For the sake of clarity, we use this word here to refer to earlier times as well.

3 For the Renaissance context of botany, see Greene 1909, 1983; Reeds 1991; Ogilvie 2006; Magnin-Gonze 2009; Bellorini 2016.

4 Del Soldato 2020, 2023.

5 Ogilvie 2006: 38.

6 Singer 1950: 174; Raven 1950: 186, 193–194; Heim 1957; Morton 1981a: ch. 6–8; Atran 1990: 58–165.

7 Magnin-Gonze 2009: 76. Some sources state 1300 rather than 1500 plants, and *De plantis* index mentions 2155 plant names.

8 The 1583 edition is the only one of Cesalpino's botanical work. No second or revised edition came out during Cesalpino's life, and, more surprisingly, no new edition was ever published after his death. This is probably because Cesalpino's work was "in advance on his time" and was understood only a century after its publication (Dughi 1957: 141). In 2022, according to the Universal Short Title Catalogue (USTC), only about seventy exemplars of the original book remain in the world, most of them in Italy.

https://doi.org/10.1515/9783111001104-001

inal text, while conveying all the concepts and nuances. The third part is a detailed commentary of the text, including necessary elements for a correct interpretation, all of which justify our translation. The fourth part provides tools to help users of the book investigate Cesalpino's work: an index of technical terms, an index of plant names, and a bibliography.

This work is the result of a close collaboration between its two authors. Corentin Tresnie is more directly responsible for the edition and translation of Cesalpino's text, and for the historical interpretation of Aristotelianism. During the translation process, Quentin Hiernaux concentrated on the botanical meaning of the phenomena described by Cesalpino, and, at a later stage, on the style of the language and on the editing process of the volume. He focused on the historical analysis of Cesalpino's botany and on the commentary. Apart from these more individualized contributions, all aspects of the work, to the smallest details of edition, translation, and interpretation, are the result of thorough discussions between the two authors.

Part I: **An Introductory Essay**

https://doi.org/10.1515/9783111001104-002

Biography of Andrea Cesalpino

Cesalpino was born in Italy, possibly in the Tuscan town of Arezzo, on 6 June 1519,[9] or, more likely, in the countryside in 1524 or 1525. He died in Rome on 23 February 1603[10] (figure 3). He studied medicine in Pisa, probably from 1543, since he graduated in 1551.[11] He studied botany with Luca Ghini (ca. 1490–1556), anatomy with Realdo Colombo (1510–1559), medicine with Guido Guidi (1509–1569), and philosophy with Simone Porzio (1496–1554). In 1557, he succeeded Ghini in botany at the Faculty of Arts in Pisa, then Francesco Violi (*fl.* mid-16th century) in medicine in 1569. From 1555 to 1583, he also succeeded Ghini as director of the Botanical Garden of Pisa,[12] founded in 1544 by Ghini with the support of the first Grand Duke of Tuscany Cosimo I de' Medici (1519–1574). The Garden is the first and oldest in the world still in existence, although its location in the city has changed several times[13] (figure 4). In 1592, Cesalpino left Pisa for Rome, disappointed by the promotion of Gerolamo Mercuriale (1530–1606) to a higher status and better paid chair of medicine than his, thanks to the support of the third Grand Duke of Tuscany Ferdinando I de' Medici (1549–1609).[14] In Rome, he taught medicine at the University La Sapienza and became archiatrist[15] to Pope Clement VIII (elected in 1592, died in 1605) after his former student, Michele Mercati (1541–1593), who introduced him to the papal court.[16]

Cesalpino is best known for his work and publications as botanist (*De plantis libri XVI*, 1583), pharmaco-therapist (*De medicamentorum facultatibus*, 1593), and physician and anatomist (*Ars medica*, 1602). However, he also dealt with philosophy (*Peripateticarum quaestionum libri quinque*, 1571)[17] and theology (*Daemonum investigatio peripatetica*, 1580). Among the natural sciences, he was interested in physics and astronomy (see especially Book III of the *Peripateticarum quaestionum libri quinque*), as well as in chemistry and geology: he wrote a treatise on metals and minerals (*De metallicis libri tres*, 1596). Some of these contributions will be mentioned below insofar as they shed light on his botanical reflections.[18]

---

9 Barnhart 1965: 1.327.

10 According to Viviani, ancient biographies relating the date of 1519 are inaccurate because they are not confirmed by the records of the baptismal register of Arezzo, and because of Cesalpino's own declarations about his age in dated official documents. Moreover, according to Viviani, Cesalpino was probably not born in Arezzo, although he lived there (Viviani 1922: 149–156).

11 Viviani 1922: 153.

12 Bellorini 2016: 67.

13 Bedini 2019.

14 Bellorini 2016: 67.

15 This title is reserved to the personal physician of the pope or a monarch.

16 Colombero 1977. On Mercati, see Pieragnoli 1853, and, more recently, Andretta 2009.

17 In *De plantis*, Cesalpino refers simply to this treatise as *Quaestiones peripateticae*.

18 For Cesalpino's biography, see also, among others: Fiorentino 1911; Viviani 1922; Mayerhöfer et al. 1959: 1.589–590; F Partington 1962: 2.89–92; Thorndike 1966: 6, chapter 40. For an integrated view of Cesalpino's multidimensional activities: Mägdefrau 1978; Strazzoni 2016.

Figure 3: Portrait of Andrea Cesalpino by G. Zocchi, engraved by G. Allegrini, 1765.

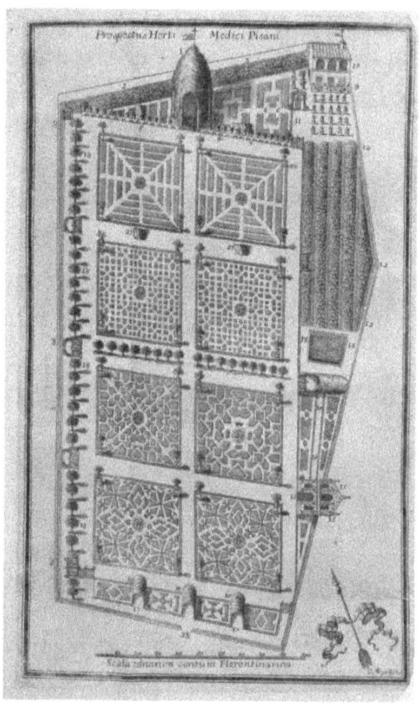

Figure 4: Plan of the botanic garden of Pisa.

Historical Context of *De plantis*

Cesalpino's contribution to the field of botany occurs at a time when medieval ideas were beginning to evolve into modern botany; his work influenced the transformation of botany during the Italian Renaissance, owing to its emancipation from, and its new relationship with, medicine.[19]

An overview of Cesalpino's predecessors is essential to understand Cesalpino's place in his times and in the field of botany, his innovations, and his legacy. The authors discussed below were selected for the innovative character of their theses or of the topic of their works. They help understand *De plantis* and the way in which it breaks with botanico-medical tradition. The points of contact between *De plantis* and the earlier botanical works presented below are not exhaustive, but illustrate the progressive emancipation of the theoretical knowledge on plants from traditional views prevalent in Cesalpino's time.[20] From the point of view of the history of science, *De plantis* marks the advent of plant morphology and systematics, through its innovative approach, unique at the time of medical botany. The discontinuity is all the clearer since Cesalpino, who was, like most botanists in his time, also a physician, deliberately chose to develop his botanical theories independently from his strictly medical work.[21]

American Edward Lee Greene (1843–1915), a botanist and historian of the field, studied the thought of Cesalpino and of many other botanists in their original text. We will frequently quote his *Landmarks of botanical history*;[22] the chapter on Cesalpino contains very relevant analyses, including extracts from *De plantis*, which is rare enough to be noteworthy. Unfortunately, this chapter is one of those left unfinished at Greene's death.[23] Alan Morton is also among the few scholars to have read the Latin text of *De plantis*, of which he provides a thorough analysis in his *History of Botanical Science*.[24] We will also mention other recent works and scholars interested in Cesalpino's *De plantis*.

Historians of science often consider that botanical science emerged in Greece with Theophrastus (circa 372–287 BCE), even though he was the recipient of a tradi-

19 Morton 1981a: 130; Bellorini 2016: 68.
20 Baldassarri 2022.
21 On the pharmacological conception of Cesalpino, see Moreau 2023.
22 Greene 1909, 1983. No fewer than 68 years separate the author's death from the edition of the second volume. During his lifetime, Greene published *Part I: Prior to 1562 A.D.* (1909), dealing with pre-modern botanists anterior to Byzantine translator of Theophrastus' work, Theodore Gaza (ca. 1398 – ca. 1478). He left *Part II* largely unfinished; it centers on Renaissance and modern botany (to Joseph Pitton de Tournefort [1656–1708]) and was edited by Frank Egerton with the help of the author's notes.
23 For the history of botany in general, see, among others: Sprengels 1817; Meyer 1854–1857; Hoeffer 1872; Sachs 1875 (Engl. transl. Garnsey 1890); Arber 1912; 1950; Davy de Virville 1954; Davy de Virville and Leroy 1969; Morton 1981a; Reeds 1991; Ogilvie 2006; Magnin-Gonze 2009; Bellorini 2016.
24 Morton 1981a.

tion now largely lost. He was indebted to his teacher Aristotle (384–322 BCE), as well as pre-Socratic philosophers Anaximander (610–546 BCE), Empedocles (490–435 BCE), and Anaxagoras (500–428 BCE). Theophrastus refers to the work of Menestor of Sybaris (*fl.* perhaps 5th century BCE), the first author to study plants specifically in the Greek Antiquity,[25] and to Diocles, a physician known for his work on food and medicinal plants conducted in the 4th century BCE. Theophrastus' observations, descriptions, and analyses of plants, recorded in *Historia plantarum*[26] and *De causis plantarum*[27], remained unsurpassed until the Renaissance.[28] A disciple of Aristotle, Theophrastus sought to expand his master's natural history of animals by applying his method and philosophical ideas to plants.[29] The importance of his work lies in the accuracy of the morphological observations underlying his classification, his innovative descriptive vocabulary, his attention to habitats, his attempts at physiological explanations, and his distinction between purely theoretical issues, and utilitarian and medical considerations. The classification of plants into four main types is one of Theophrastus' major contributions to the field:[30] trees, shrubs, undershrubs, and herbs. Cesalpino adopts this quadripartition in *De plantis*, while developing the most important theoretical concepts at the foundation of so-called modern botany. Along with Aristotle, Theophrastus was Cesalpino's main source.[31] A treatise *De plantis* was for a long time attributed to Aristotle but is now attributed to Nicolaus Damascenus (1st century BCE)[32]. This text was copied and commented on extensively in the Middle Ages, in the Arabic World and in the West, by

---

25 Although Menestor is considered as the first botanist, there is no record of his writings, and what little is known of them is discussed by Theophrastus.

26 Theophrastus, *Historia plantarum*. (ed. and Engl. transl. Hort 1916; ed. and Fr. transl. Amigues 1988–2006).

27 Theophrastus, *De causis plantarum* (ed. and Engl. transl. Einarson-Link 1976–1990; ed. and Fr. transl. Amigues 2012–2017).

28 Magnin-Gonze 2009: 11–31.

29 Amigues 2001: 72–74.

30 Theophrastus' four plant types (εἶδος) (*Historia plantarum* I, 3, 1 [ed. and Fr. transl. Amigues 1988–2006: 1.9–10]), sometimes called the four main genera, are not to be understood in their modern taxonomic sense, but in the Aristotelian perspective of grouping according to shared essential properties. When Renaissance authors, and Cesalpino, refer to plant genera, the term must be understood as referring to any grouping above species. Cesalpino's use of genus can therefore cover the current usage (i.e., the taxon strictly superior to related species), but can also refer to what we would today call family taxa, order taxa, etc., or to groups without current taxonomic value (e.g., species grouped according to their living environment, etc.).

31 Theophrastus is one of the few authors explicitly mentioned in *De plantis*, in the Dedication letter: IV; and in the text itself: ch. 12: §134.

32 Nicolaus Damascenus [Pseudo-Aristotle], *De plantis* (ed. and Engl. transl. Drossaart Lulofs and Poortman 1989; ed. and Fr. transl. Federspiel and Cronier 2018). The identity of the author remains uncertain. The English edition and translation mention Nicolaus Damascenus as the author of the text but the French edition and translation attribute it to Pseudo-Aristotle.

figures such as Albert the Great (1193–1280) in his works on plants.[33] Julius Caesar Scaliger (1484–1558) was the first to reject Aristotle's authorship of *De plantis*,[34] but it continued to circulate among Renaissance thinkers, albeit with less influence, and remained a source for Aristotelian botanists like Cesalpino.[35]

Theophrastus' treatises were soon lost in Byzantium and the West. From the early Middle Ages throughout the medieval period, botany does not seem to have existed as an independent discipline, but rather as a subdivision of medicine and therapeutics with very practical applications. Plants were described concisely, and studied for their therapeutic properties, and their uses, including as remedies. Theoretical study of the forms and functions of plants did not exist as such in medical botany. Species were often listed in alphabetical order or according to the properties of the plants. Medieval knowledge of plants was founded in the medical writings belonging to the corpus attributed to Hippocrates (circa 460 – between 375 and 351 BCE), as well as in the works of Crateuas (2nd–1st century BCE), Dioscorides[36], Pliny the Elder (23/24–79 CE), Galen (circa 129 – after 216 CE [?]), and, to a lesser extent, of the Latin agronomists: Cato the Elder (234–149 BCE), Varro (116–27 BCE), Columella (4–70 CE), and Palladius (*fl.* 371–395 CE).[37]

From the 10th century onwards, the Middle East played an important role in the conservation and transmission of texts. Physicians like Avicenna (980–1037) studied plants unknown to earlier scientists. Avicenna also developed new theories on vegetative faculties and plant life.[38] In the 12th and 13th centuries, botany began its slow emancipation from medicine. Observation and experimentation were encouraged by the creation of universities, and by the development of nominalism, a philosophical view prioritizing empirical reality over concepts. However, the teaching of botany in universities focused on medical aspects, and botany was regarded as a science ancillary to medicine. Albert the Great nevertheless recorded new morphological and anatomical details and offered original descriptions of plants in his *De vegetabilibus*.[39] Meanwhile, popular and practical knowledge of plants found an echo in religious communities, as illustrated by the example of Hildegard of Bingen (1098–1179).[40]

Like the Renaissance of the arts and of humanism, the Renaissance of botany, in which Cesalpino was involved, began in Europe in the 15th century and flourished par-

33 Arber 1912: 3–5; Panarelli 2020.

34 Scaliger 1556.

35 On the complex reception of *De Pantis* in early modern science, see Reeds 1991: 9; Repici 2005; Baldassarri 2020; Hiernaux and Tresnie 2023a.

36 Dioscorides, *De Materia medica* (ed. Wellmann 1906–1914).

37 Nisard 1864.

38 Fatigati 2021; according to Tawara 2014, Avicenna reached the conclusion that plants could not be alive.

39 Albert the Great, *De vegetabilibus* (ed. Meyer, 1867).

40 Magnin-Gonze 2009: 29–46.

ticularly rapidly in Italy,[41] a country which enjoyed economic prosperity at that time. As a result, universities in many Italian cities began to create botanical gardens in the 16th century. Exploration of remote lands brought to Europe a multitude of new plant species. The development of printing promoted the dissemination of knowledge and illustrations, and the technique of woodcutting, and, later, of copperplate engraving, enabled the inclusion of life-like representations in printed botanical books.[42] Representations of plants were also transformed by the development of realist trends in painting.[43] In philosophy, the Aristotelian doctrine continued to prevail and to influence science, as it had done during the Middle Ages. Moreover, several Italian physicians distanced themselves from Dioscorides and Pliny, the successive copies of whose works contained errors.[44] They were helped by the rediscovery of the work of Theophrastus, translated into Latin by Theodore Gaza (circa 1400 – circa 1478) between 1451 and 1454, and printed in 1483.[45] Cesalpino probably owned these translations, since he mentions Theodore Gaza[46]. For the first time, the comparative study of living plants and the specialization of the relevant vocabulary began to challenge the authority of previous texts.[47]

The relationship between Cesalpino and his predecessors and contemporaries is rarely explicit. A typical Renaissance botanist and intellectual, Cesalpino mentions ancient thinkers, mostly physicians, whose texts he has most certainly read or commented, but he hardly mentions his contemporaries. Cesalpino's sources also remain implicit or cryptic in his philosophical works.[48] Cesalpino was deeply Aristotelian, and Renaissance botany relied heavily on the authority of ancient texts. According to a non-exhaustive inventory established by Greene,[49] Cesalpino mentions the following names in *De plantis*:[50] Theophrastus, Columella (4–70 CE), Ermolao Barbaro (1454–1493), Aëtius [of Amida] (6th century CE), Matthaeus Sylvaticus (ca. 1280 – ca. 1342), Pliny [the elder], Galen, Dioscorides, Serapion the Younger [Ibn-Sarabiyun] (12th/13th century),

41 For the revival of botany in 16th-century Germany, and in present Belgium and Netherlands, see Greene 1909.

42 Arber 1912; Reeds 1991.

43 O'Malley and Meyers 2008.

44 Touwaide 1984.

45 Theodore Gaza also translated Aristotle's zoological treatises. Although Theodore Gaza was not a botanist, his translation of Theophrastus is of high quality and constitutes a considerable advance for this science. Editions of Theophrastus were based on the corrected reproduction of Theodore Gaza's text until the publication of Hort's translation in 1916 (Greene 1983: 2.524–525). On Gaza see Greene 1983: 2.518–528.

46 Cesalpino 1583: 36; 37; 97 etc. Theodorus (Gaza) is mentioned more than eighty times in *De plantis*.

47 Touwaide 2016 and 2017.

48 Del Soldato 2022.

49 Greene 1983: 2.1021n64.

50 For scholars whose names are cited in abbreviated form in *De plantis* and by Greene, we provide in square brackets the full surnames currently in use.

Paul of Aegina (7th century), Solinus [Gaius Julius Solinus] (3rd or 4th century), Ludovicus [Pontus] Romanus (1409–1439), Athenaeus [of Attalia] (1st century CE), Mesuë [Yuhanna ibn Masawaih] (ca. 777–857), Castor Durante [Da Gualdo] (1529–1590), and Garcia d'Orta [Garcia del Huerta] (1501–1568).[51] In addition to Theodore Gaza, Barbaro, Durante, Ludovicus Pontus Romanus and d'Orta, who are Renaissance authors, Cesalpino cites a few other Italian contemporaries who contributed to the emergence of botanical science in a modern sense.

Among the precursors of the Italian Renaissance of botany, before Cesalpino and after Theodore Gaza, we should mention Niccolò Leoniceno (1428–1524), Marcello Virgilio Adriani (1464–1521), Giovanni Manardi (1462–1536), Antonius Musa Brasavola (1500–1555), Luca Ghini, Luigi Anguillara (ca. 1512–1570), Pietro Andrea Mattioli (1501–1578) and Fabio Colonna (1567–1650), a contemporary of Cesalpino. Their contributions to the transformation of botany and their possible influence on Cesalpino are outlined below.[52]

A revival of botany also took place in Western Europe (Germany, France, Flanders, Holland and Switzerland) at that time, prompted by Otto Brunfels (ca. 1488–1534), Valerius Cordus (1515–1544), Hieronymus Bock, known as Tragus (1498–1554), Matthias de L'Obel [Lobelius] (1538–1616), Rembert Dodoens [Dodonaeus] (1517–1585), Charles de l'Écluse, known as Clusius (1526–1609), Jean Ruel [Ruellius] (1474–1537), Conrad Gessner (1516–1565), Leonhart Fuchs (1501–1566) and brothers Jean (1541–1612) and Gaspard Bauhin (1560–1624). Our study will focus on the Italian authors, because Cesalpino's Western European predecessors and contemporaries had but little influence on *De plantis*.[53] Although the thoughts and works of Italian botanists and physicians of the Renaissance constitute a unique context for Cesalpino, it is difficult to determine his exact sources and influences beyond a few explicit references in *De plantis* to Theodore Gaza, Barbaro, Mattioli, Ghini, Jean Ruel (one of the rare mentions of a non-Italian scholar), Antonius Musa [Brasavola], Anguillara (mentioned as Aloysius Anguillara),[54] and possibly Adriani.[55] These authors present links, either filial, epistolary, institutional,

51 Greene (1983: 2.1021n64) indexes "Don Garzias medicus" separately, but this heading most likely also refers to Garcia d'Orta, since Cesalpino mentions him several times in connection with plants found in India. Indeed, d'Orta is famous for his treatise *Colóquios dos simples e drogas da India* published in Portuguese in 1563, translated into Latin in 1572, and into Italian in 1575. Cesalpino was certainly aware of this treatise. Greene also mentions a "Garzias Lusitanus", a name which does not correspond to any patronymic retained by history. It is therefore either Garcia d'Orta or, more likely, Amatus Lusitanus [João Rodrigues de Castelo Branco] (1511–1568), another famous Portuguese physician who lived and taught in Italy at the time of Cesalpino. On Lusitanus, see Findlen 1999.

52 On the authorities mentioned by Cesalpino, see Heideklang 2023.

53 Sachs 1875 (Engl. transl. Garnsey 1890: 40).

54 Luigi (or Aloysius) Anguillara served as a pseudonym for Luigi Squalermo, his real name.

55 Cesalpino mentions Marcellus: 99 and Adriani: 7. These could be references to Marcello Virgillo Adriani.

or cultural, with other authors such as Leoniceno and Manardi, whom we can therefore plausibly include in Cesalpino's intellectual landscape, even if indirectly.

Leoniceno[56] is famously the first author to correct Pliny's Latin text. His study of the Greek text of Dioscorides, to which many of Pliny's passages were similar,[57] and his own botanical knowledge led him to notice inconsistencies in the descriptions or identifications of plants, and to detect translation errors. Although Leoniceno's concern was primarily medical when he published *Plinii ac plurium aliorum auctorum qui de simplicibus medicaminibus scripserunt errores notati*[58] (*Indication of Errors in Pliny and in Several other Authors who have written on Medicinal Simples*) in 1492, the philosophical scope of his critical work is much broader, and includes considerations on epistemology and on the scientific method:[59]

> The cause to which Leoniceno devotes himself here is not alone that of the betterment of botany and the materia medica. The force of his argument has broader bearings. Its scope is as wide as the whole field of science and transmitted knowledge of every kind. He is appealing for the acceptance of demonstrable truth, against abject and credulous veneration of authority. He is erecting a landmark in the advancement of all knowledge.[60]

Leoniceno defends an empiricism of the senses inspired by Aristotelianism and develops epistemological reflections on the personal acquisition of knowledge.[61] He sought knowledge of the things themselves, rather than of words, and he suggested the use of unambiguous terms to refer to, and identify a single thing.[62] His ideas are almost Cartesian *avant la lettre*, but his position was controversial at the time, and several of his contemporaries, such as Pandolfo Collenuccio (ca. 1444–1504), were staunch defenders of Pliny.[63] Leoniceno edited or contributed to the edition of the works of Aristotle (5 volumes, 1495–1498), Dioscorides (1499), and Galen (published posthumously in 1525). Leoniceno probably taught Ghini[64] – Cesalpino's mentor.

Cesalpino does not mention Leoniceno directly, but he quotes Ermolao Barbaro, who supported the critical reading of Pliny in his *Castigationes plinianae* (1492–1493)[65] published at the same time as Leoniceno's work, although independently. Unlike Leoniceno, Barbaro did not directly challenge the authority of the ancient author, but ascribed

56 On Leoniceno, see Greene 1983: 2.528–543; Touwaide 2000 and 2007; Ogilvie 2006: 30–32, 126–133.
57 Pliny used the same sources as Dioscorides.
58 Leoniceno 1492.
59 Edwards 1976; Touwaide 2007.
60 Greene 1983: 2.539.
61 Leoniceno 1492, see Greene 1983: 2.542–543.
62 Touwaide 2000, 2007, 2017.
63 On Collenuccio, see Greene 1983: 2.544–552.
64 Bellorini 2016: 62.
65 Barbaro 1492.

errors to poor copies and translations.[66] He also produced the first post-medieval Latin translation of Dioscorides, published posthumously in 1516[67] with an appendix entitled *Corollarium Dioscorides*, in which he addressed pharmaco-therapeutical and practical issues through an update of all ancient Greek and Latin knowledge of the plants cited by Dioscorides.[68] According to Greene,[69] Barbaro was the first botanist to supplement ancient treatises with his own field observations.[70]

In 1518, Marcello Virgilio Adriani, known as the Florentine Dioscorides,[71] also published a Latin translation of Dioscorides. In his commentary, he rejected the taxonomic relevance of color and proposed, for the first time, a rearrangement of Dioscorides' genera, which he considered particularly artificial. He spearheaded the development of a natural classification according to genera affinities, although Cesalpino was the first to undertake explicitly and systematically such a task. Marcello Virgilo Adriani saw the necessity to clarify nomenclature:

> Nullibi similitudine nominum decaepti grauius quam in medicina peccamus: Interest enim uitae et sanitatis humanae.
>
> Nowhere do men so gravely err in medicine, as where they allow themselves to be deceived by similitude in names; for here human life and health may be at stake.[72]

Cesalpino's botany takes this issue very seriously indeed, since he concentrates on the observation of similarities between plant parts, rather than on the artificial resemblances between their names.[73]

Giovanni Manardi contributed to the transformation of the relationship between medicine and botany during the Renaissance.[74] He eliminated certain plants from pharmacotherapy on phytological grounds: their botanical descriptions did not correspond to those in ancient texts, nor to the plants sold by apothecaries. Long before central European botanists such as Tragus[75] or Brunfels, Manardi introduced descriptions of plants previously lacking it (notably the lily of the valley [*Convallaria majalis* L.]),

---

66 On the relationship between Leoniceno and Barbaro on Pliny, see Touwaide 2000.
67 Barbaro 1516.
68 Ramminger 2005.
69 Greene 1983: 2.563–565.
70 On Barbaro's botany, see also Greene 1983: 2.553–568; Reeds 1991: 11, 19–21, 27, 106, 169; Bellorini 2016: 62–63.
71 This author is best known for his literary work and his political career as a diplomat. He was the professor of Niccolò Machiavelli (1469–1527). On Adriani's botany, see Greene 1983: 2.569–583.
72 Adriani 1518: 1.7, quoted and translated by Greene 1983: 2.582.
73 Cesalpino criticizes, for example, the grouping of species under the same name because of their common use, whether medicinal (ch. 12: §123), dietary or, more broadly, utilitarian (ch. 12: §127).
74 On the relationship with Leoniceno's work, see Mugnai Carrara 1994.
75 On Tragus, see Greene 1909: 220–262; Arber 1912: 50–58.

whereas his predecessors merely tried to match each encountered plant to one of the ancient descriptions. Thus, Manardi may be considered as a precursor of Frenchman Jean Ruel, one of the only contemporary non-Italian botanists quoted by Cesalpino – although he does not mention Manardi.[76]

Ruel wrote one of the first post-medieval treatises wholly devoted to botany, independently from pharmacotherapeutics and other utilitarian considerations. He also initiated field botany and published in 1516 a translation of Dioscorides,[77] the most successful of the three produced at that same time. His major contribution was *De Natura stirpium libri tres* (1536),[78] in which he provided exhaustive and brief descriptions of all the plants that he mentioned. Ruel's name is associated with great terminological rigor and lexical innovations; he favored description over illustration. His work is comparable to the later publications by German author Valerius Cordus (1515–1544).[79] Ruel's contributions to the analysis of the morphology of flowers are significant, especially since Cesalpino too relied on floral morphology and its descriptions in his classification. Ruel distinguished corolla flowers from other inflorescences and differentiated the perigynous or hypogynous implantation of the flower from the fruit (the ovary of which is consequently superior or inferior). Moreover, he was the first to identify several parts in the flower itself, and to describe them. Cesalpino adopted these distinctions and descriptions as criteria for classification, which was a novelty.[80] In his dedication of *De plantis* to the Grand Duke of Tuscany, he explained that he attempted to develop a new method of rational classification, whereas his contemporary Ruel had only sketched it:

> Apud nostros autem Ruellius tentauit quidem, sed preaeter ea, quae a Theophrasto excerpsit circa rationem communem, vlterius nequaquam est progressus.
>
> Closer to us, Ruellius made some attempts, but he made no progress further than what he took from Theophrastus concerning the common notion.[81]

Although Ruel distinguished certain natural genera, "these groupings are made more in deference to the economic values than any other consideration",[82] a choice openly criticized by Cesalpino in his own classification project:

> Nam praeter frumenta et legumina, quae communi nomine Fruges vocantur, et praeter Olera vix alia reperiemus manifesta, et haec ab usu potius assumpta, quam a formae similitudine, quam

76 On Manardi's botany, see Greene 1983: 2.583–597.
77 Ruel 1516.
78 Ruel 1536.
79 On Valerius Cordus, see Greene 1909: 270–314; Arber 1912: 65–67; Sprague 1939; Reeds 1991: 13, 18, 23, 77, 125, 143; Ogilvie 2006: 145–148; Magnin-Gonze 2009: 68–70.
80 Cesalpino 1583: ch. 14: §152.
81 Cesalpino 1583: Dedication letter IV.
82 Greene 1983: 2.641.

quaerimus: inter Fruges enim Sesamum numerant, et Irionem; quia eorum seminibus vescimur, formis maxime a caeteris diuersa. In Oleribus autem longe plura reperiemus formis distantissima, vt Asparagum, Brassicam, Portulacam, Cicorium, Bulbos, aliaque multa, quae in nulla re alia conueniunt praeter quam in vsu cibario, vel tenera germina, vel folia, vel radices, vel totas plantas nobis assumentibus.

Furthermore, we will discover that cereals and legumes, for which we use the common term "crops", as well as vegetable varieties, are quite different, and that these genera are categorized together more as a result of their use than because of their similitude in form, which is what we are studying here. Indeed, included in crops we have sesame and hedge mustard, because we nourish ourselves from their seeds, even though their form is very different from those of other crops. In the vegetable varieties, we find even more major differences and many other [species] which conform in no other way but their use in food; [we see that] by looking sometimes at the tender shoots, sometimes at the leaves, sometimes at the roots, sometimes at the whole plants.[83]

Moreover, in plant physiology, Ruel offered a compilation of the ideas from ancient text without any additions, whereas Cesalpino, even though he remained influenced by Antiquity, proposed his own reflections on the circulation of sap, nutrition, and growth, for example. Since Cesalpino quoted "Ruellius" several times in *De plantis* and drew in his preface an explicit (and therefore political) contrast between his approach and Ruel's, his knowledge of Ruel's work must have been more than superficial.[84]

In 1536, the year of Ruel's major publication, Brasavola published his *Examen omnium simplicium medicamentorum, quorum in officinis usus est*,[85] an attempt to identify Dioscorides' plant species, and to determine whether Dioscorides' descriptions corresponded to plants bearing the same names in Brasavola's time. Brasavola described only a few plants that do not seem to correspond to any ancient description. Moreover, he described and grouped together three related genera (*Fragaria*, *Tormentilla*, and *Potentilla*) on account of the morphology of their fruit; this anticipates the generalization of this mode of classification by Cesalpino.[86] Brasavola's treatise was most significant in the field of pharmacotherapy: he criticizes medieval pharmacopoeia, the incoherence of which was increasingly recognized from this period onwards. Brasavola is also noteworthy for his experimental approach, and for asserting the independence of botany from medicine.[87]

Luca Ghini may be considered as one of the most influential figures for Cesalpino as a botanist.[88] Cesalpino was one of his disciples, as were Ulisse Aldrovandi (1522–1605), Luigi Anguillara, Bartolomeo Maranta (1500–1571), and John Falconer (active in

83 Ch. 12: §127, see also §123 and §135.

84 On Ruel's botany, see Greene 1983: 2. 598–657.

85 Brasavloa 1536.

86 Greene 1983: 2.681.

87 Magnin-Gonze 2009: 81. On Brasavola's botany, see also Greene 1983: 2.658–701; Magnin-Gonze 2009: 48, 81; Bellorini 2016: 70.

88 On Ghini, see Sachs 1875 (Engl. transl. Garnsey 1890: 18); Greene 1983: 2.702–721; Magnin-Gonze 2009: 54–57; Thiers 2020: 14–22.

the 1540s). Ghini, a physician and botanist, taught in Bologna and Pisa among the most renowned of his time. He founded the Botanical Garden of Pisa in 1544 under Cosimo I, Grand Duke of Tuscany, and served as its director until Cesalpino succeeded him in 1555. He also created the Botanical Garden of Florence in December 1545. The creation of botanical gardens enabled field studies and promoted an empirical approach to botany. Ghini garnered famed among his contemporaries, with whom he corresponded, but did not publish much of his work, and almost none of his writings, apart from his letters, has survived to the present day.[89] As a result, the historiography of the personage and the affirmation of his influence on botany remain speculative. And yet, it is reasonable to consider Ghini as a pioneer of accurate botanical descriptions, owing to his scientific sense of observation and his adoption of a method of morphological classification of plants that did not rely on utilitarian considerations,[90] at a time when this strict approach and the alphabetical order inherited from the Middle Ages prevailed.[91] Ghini had intended to write his own herbarium, but instead shared his notes and observations with Pietro Andrea Mattioli (1501–1578), whose herbarium was already more advanced. Through his botanical work, Cesalpino fulfilled the theoretical and practical projects of his master, and Ghini largely achieved posterity through the writings of Cesalpino and Mattioli.[92] Ghini is also regarded as the inventor of the technique of the dried herbarium, which he passed on to his disciples[93] – Cesalpino's and Aldrovandi's collections of dried plants are among the oldest still in existence.[94] Ghini's correspondence with Mattioli reveals that Ghini regularly herborized and sent specimens to Mattioli, although they may not have been dried and pressed between sheets of paper;[95] as such, the invention of the dried herbarium is sometimes attributed to Englishman John Falconer, known to have dried plants between sheets of paper for conservation in a herbarium.[96] In any case, there is no mention of drying plants between pages in Antiquity or before the above-mentioned authors. Epistemologically, Ghini's interest in real specimens, dried or not, for the study of botany is unmistakable; in his understand-

89 Only comments on specific plants written for Mattioli were preserved in manuscripts and published as *Placiti* (Morton 1981a: 156n40; Bellorini 2016: 60–67). Some lectures on *Materia medica*, transcribed by Aldrovandi, and a medical treatise about syphilis (*De morbo Gallico*) transcribed by a student also remain (Bellorini 2016: 60). For the edition of these texts, see Toni 1907; Sabbatani 1921.

90 The correspondence between Gessner and Ghini indicates that the latter had already given thought to diagnostic features and classification (Magnin-Gonze 2009: 74). On Gessner, see Arber 1912: 90–93; Greene 1983: 2.747–797; Ogilvie 2006: 34–38, 236–240; Magnin-Gonze 2009: 74–79.

91 Even Leonhart Fuchs and Jean Ruel, despite their descriptive contributions, arrange plants in alphabetical order.

92 On Mattioli, see Arber 1912: 79–85, 186–190; Greene 1983: 2.798–806; Ferri 1997; Reeds 1991: 155–164; Magnin-Gonze 2009: 60–61.

93 Ogilvie 2006: 165.

94 On Aldrovandi, see Reeds 1991: 17–18, 34–36, 56, 109, 113; Magnin-Gonze 2009: 55–57.

95 Saint-Lager 1885.

96 Saint-Lager 1885: 18–25.

ing and teaching of botany, he prioritized the empirical plant – living or dead – over its description or representation.[97] Herbaria prompted international advances in botanical research through the exchange of specimens, and, above all, allowed botanists to study plants in detail, and to teach botany at any time of the year. The same was true of botanical gardens, which multiplied throughout Italy and then Europe from the late 16th century into the 17th century.

Anguillara was, like Cesalpino, a disciple of Ghini. He shared his master's interest in herbalism and field study and founded the second oldest botanical garden in Padua in 1545. This garden is the oldest to have been preserved in its original location. Anguillara became famous with the publication of his treatise *Semplici*,[98] written in Italian, on the medicinal and dietetic plants of the Ancients. Anguillara wished to identify in their native environment all the plants mentioned by Dioscorides and Theophrastus, and therefore spent much of his life travelling in Greece, around the Mediterranean, and in the Near East.[99]

Following the iconographic advances achieved by Otto Brunfels and Fuchs in Germany,[100] Italian botanist Pietro Andrea Mattioli proposed illustrations of plants in a treatise entitled *Petri Andreae Matthioli medici senensis commentarii, in libros sex pedacii Dioscoridis anazarbei, de materia medica, adjectis quàm plurimis plantarum & animalium imaginibus, eodem authore* (1554)[101], a revised edition of his Italian translation of *De materia medica* entitled *Comentarii in sex libros pedacii Dioscoridis de materia medica* (1544)[102]. Mattioli's original descriptions surpassed his predecessors', and his commentary on Dioscorides, supplemented by his own observations, resulted in sixty editions in many languages.[103]

Fabio Colonna is a contemporary, albeit younger, of Cesalpino. His first botanical treatise, the *Phytobasanos*[104] was published in 1592, and investigated the medicinal plants used by apothecaries by comparing them with living individuals in their original environment and at different stages of their life cycle. He shared with Cesalpino his critical approach to popular ideas about plants, and his refusal to trust ancient authorities unquestioningly. Colonna described ninety new species, and introduced copperplate engraving in botany, more accurate than the woodcuts used in the 16th century, for example in Otto Brunfels' *Herbarium vivae eicones* published between 1530 and 1536)[105].

---

97 Bellorini 2016: 60–67.
98 Anguillara 1561.
99 On Anguillara: Greene 1983: 2.723–746; Reeds 1991: 109, 159; Magnin-Gonze 2009: 55–56.
100 On Fuchs, see Greene 1909: 192–219; Arber 1912: 58–64, 175–185; Ogilvie 2006: 34–36, 113–114, 194–197; Magnin-Gonze 2009: 60–68. On Brunfels, see our section on the status of illustrations.
101 Mattioli 1554.
102 Mattioli 1544.
103 Magnin-Gonze 2009: 60.
104 Colonna 1592.
105 Brunfels 1530–1536.

He rejected classifications based on leaves, adopted by authors such as de L'Obel, and focused instead on flowers and fruits, as Cesalpino recommended.[106]

These Renaissance scholars, exclusively men,[107] were above all philologists studying and translating ancient botanical treatises, but they would not have achieved rigorous translations and corrections without a thorough knowledge of plants.[108] The broad outlines of their work combined with ancient knowledge allow us to understand Cesalpino's intellectual achievement and to appreciate the scope of his innovations, influence, and limitations.

Outline of Book I

Book I of *De plantis*, translated and commented in this volume, is the most important among the sixteen books of Cesalpino's treatise. As famous German botanist and historian of his discipline Julius von Sachs (1832–1897)[109] writes in his *History of Botany*, *De plantis* marks the origin of descriptive botany:

> If the value of the contemporary German botanists lies pre-eminently in the accumulation of descriptions of individual plants, and these, it is true, occupy fifteen books of Cesalpino's work, it is on the contrary the introduction in the first book, a discussion of the general theory of the subject, which in his case is of much the higher importance for the history of botany.[110]

Indeed, Book I sets out Cesalpino's theoretical principles, his methodology, his general observations on plant morphology and anatomy, and his physiological hypotheses and explanations. It also provides the principles underlying his method of classification and his groupings of species into more general genera. Book I also reveals Cesalpino's intellectual influences: Theophrastus', whose work Cesalpino knew and whose division of plants into four major botanical groups he adopted, but also, and above all, Aristotle's, which permeates all of Cesalpino's work. Book I is preceded by a letter dedicated to

106 On Colonna's botany, see Greene 1983: 2.832–846.

107 There is no evidence of female botanists publishing their work during the Renaissance. However, Ghini's correspondence reveals that he regularly exchanged with women about plants.

108 Greene 1983: 2.611–612; Bellorini 2016: 63.

109 Julius von Sachs is considered as the leader of experimental plant physiology in Germany, and as the promoter of laboratory methods in the study of plants in Europe in the second half of the 19th century. Charles Darwin's son Francis studied plant physiology with Sachs in Würzburg before he helped his father to study plant growth and movement. These studies resulted in a heated controversy between Charles and Francis Darwin, and Sachs (see Bernier 2013).

110 "Liegt der Werth der gleichzeitigen deutschen Kräuterbücher ganz vormiegend in der Unhäufung zahlreicher Einzelfbeschreibungen, die zwar auch in diezem Wert 15 Bücher füllen, so ist bagegen für die Geschichte der Botanik in diesem Falle die algemein theoretische Einleitung im ersten Buch von ganz hervorragenber Bedeutung." (Sachs 1875: 45; Engl. transl. Garnsey 1890: 42).

Grand Duke Francesco I of Tuscany (1541–1587) and included as a preface to *De plantis*. This letter contains important methodological elements and clarifies Cesalpino's position towards some of his predecessors and contemporaries. Four laudatory poems, an index, and an erratum complete the text.[111]

Book I is divided into fourteen chapters offering a general discussion of the nature of plants, their vital functions, and their parts (table 1). Each chapter has a specific topic, and all follow in a logical order. The first book can be divided into two parts. Chapters 1 to 11 focus on morphology, anatomy, and physiology, and describe the essential operations of plants; chapters 12 to 14 introduce the classification into genera and species, and the methodological principles followed in the other fifteen books.

Chapter 1 discusses the soul of plants, its location, the main parts of the plant in general and the faculties with which each is associated. Chapter 2 describes the functioning of the nutritive faculty for obtaining and processing food. Chapter 3 deals with growth, the other aspect of the nutritive faculty, and the role of the bark, pith, and leaves. Chapter 4 focuses on the early stages of plant growth, i.e., development from seed or sobole. This provides a transition to the analysis of the other main faculty of plants, namely reproduction, which is the focus of Chapter 5. Chapter 6 concentrates on seeds, their morphology, germination, and mode of division. Chapter 7 is devoted to flowers, their composition, morphology, and their role in the protection and reproduction of the plant. Chapter 8 describes the fruits, the pericarp, and the barks that protect the seed. The genesis, usefulness, and composition of the pericarp are discussed in chapter 9, and other forms of protection in chapter 10, which also contains a section on fruiting. Finally, chapter 11 deals briefly with secondary parts such as tendrils, thorns, or moss.

Chapter 12 opens the second part of Book I by reviewing the four traditional genera of Theophrastus and discussing the criteria according to which subgenera[112] are usually defined. Chapter 13 argues for a definition of genera and subgenera based solely on the essential parts of the plant (root and stem) and the main operations (nutrition and reproduction) that its vegetative soul enables it to perform *via* these parts. Chapter 14 extends this argument by examining specific problems arising when subgenera are defined according to the reproductive operation. These chapters lay the theoretical background underlying the structure of plant descriptions in the following fifteen books. It is worth noting again that almost no practical or medical considerations are included in this first book, a small revolution for the time.

111 On this paratext of *De plantis*, see Heideklang 2023.

112 In the context of *De plantis* "genera" refers indistinctly to higher groups like the four main types (εἶδος) of Theophrastus – trees, shrubs, undershrubs, and herbs (*Historia plantarum* I, 3, 1 [ed. and Fr. transl. Amigues 1988–2006: 1.9–10; ed. and Engl. transl. Hort 1916–1928: 1.23–25]) – or to groups at the level of families in the current taxonomic sense. We call these groups "subgenera", to clarify Cesalpino's choice to divide the first four main genera (in chapters 12 and 13). "Subgenera" must not be considered here as a subgenus in the current taxonomic vocabulary, which designates the taxon under the current genus and higher than the species.

Books II to XVI, which are not translated here, contain descriptions of about 1500 species. Cesalpino's innovation in these books resides in the sequence in which the species are presented: instead of ordering them according to their properties or to pharmacotherapeutic considerations, Cesalpino follows the four major Theophrastean genera: books II and III are devoted to trees and shrubs, books IV to XVI to undershrubs and herbs. The morphological criteria and the classification method proposed in Book I are applied in the organization of the fifteen books, but not in their internal structure: Cesalpino presents one by one numerous species of plants and specifies their morphological properties and their various uses – including their medicinal virtues – alongside the descriptions of these species by authors such as Theophrastus, Pliny, or Dioscorides.

Cesalpino's Method, Intellectual Innovations, and Influence

Cesalpino inherited the epistemological innovations of his predecessors, not as a physician anxious to correct the errors of the Ancients, but as an advocate of a new working method based on personal reflection, observation, and the available knowledge, rather than on previous treatises.

> His work [...] bears the unmistakable imprint of a highly original mind and of the most discerning powers of observation.[113]

According to Morton, Cesalpino was responsible for the first great advance in botany since Theophrastus, and botanist Ernst Meyer (1791–1858) refers to him as "unquestionably the greatest botanist of his century" in his *Geschichte der Botanik*.[114]

Book I of *De plantis* contains no explicit reference to Pliny and Dioscorides, and borrows Theophrastus' ideas only insofar as Cesalpino found them worthy of consideration – when he did not, he did not hesitate to criticize them. The relationship between science and philosophy was complex in the 16th century, and to assess the quality of the scientific knowledge of the time solely in terms of its emancipation from Aristotle is both too restrictive, and anachronistic. However, this approach prevailed for a long time, for example in these scathing lines from the *Dictionnaire encyclopédique des sciences médicales*:

> Cesalpino was one of those superior geniuses whose penetration overcomes the greatest difficulties. But he was too slavishly attached to Aristotle's doctrine, which he warmly defended against Galen's, adopted by the schools of that time. His philosophical writings breathe only Aristotelian

113 Morton 1981a: 128.
114 Morton 1981a: 141.

> doctrine, expressed, it must be said, in a twisted, diffuse manner, which makes them very difficult to understand, often presenting nothing but an inextricable web of empty words that one may interpret as they wish. Also, to avoid an extraordinarily tiring reading, which would defeat the most tenacious mind, amateurs of ancient philosophy are advised to set aside the text of the Italian doctor.[115]

This very hasty assessment, reproduced in reference works,[116] certainly hindered the reading and translation of Cesalpino's Latin texts, even though these same works generally recognize the value of Cesalpino's botanical contribution.[117] It is true that Cesalpino's writing is complicated and sometimes elliptical, and requires a proper interpretation.[118] However, the abilities of Cesalpino's philosophical background is crucial for understanding his botany, and should no longer be considered as an independent facet of his work. According to American historian of science and philosophy Charles Schmitt (1933–1986), Cesalpino was the most original thinker in the field of philosophy at the University of Pisa during Galileo's time.[119] Cesalpino's original philosophical thinking likely prompted him to adopt a new approach to the study of plants; this triggered the emergence of botany as an independent discipline.[120] In the preface of *De plantis*, Cesalpino himself wrote that the medical orientation of the study of plants and the absence of philosophy in the works of naturalists were the reasons why the problem of classification had not yet been solved.

Beyond naturalistic considerations, Book I of *De plantis* is a treatise on natural philosophy in its own right. The Aristotelian influence is noticeable, especially in the theory of the soul's faculties; yet, although Aristotelianism is fundamental for Cesalpi-

115 "Cesalpino était un de ces génies supérieurs dont la pénétration surmonte les plus grandes difficultés. Mais il fut trop servilement attaché à la doctrine d'Aristote, qu'il défendit avec chaleur contre celle de Galien, adoptée par les écoles de ce temps-là. Ses écrits philosophiques ne respirent que la doctrine aristotélique exprimée, il faut bien le dire, d'une manière entortillée, diffuse, qui les rend très difficiles à comprendre, ne présentant souvent qu'un tissu inextricable de mots vides de sens que chacun peut interpréter comme il l'entend. Aussi, pour éviter une lecture extraordinairement fatigante, devant laquelle l'esprit le plus tenace se trouve vaincu, peut on conseiller aux amateurs de l'ancienne philosophie de laisser là le texte même du médecin italien." (Chéreau 1873: 543–545, our translation).

116 See also Harvey-Gibson 1919: 19–21.

117 Several historians of botany, notably Hoeffer (1872), did not hesitate to declare, in a historicist and positivist perspective, that Cesalpino was the only Renaissance botanist to produce work worthy of interest. This is obviously an exaggeration.

118 Morton is more enthusiastic regarding the "greatest scientific and historical significance" of the work "written in clear but condensed philosophical style" adding that "its importance has not always been fully appreciated, in spite of the very hight tribute paid to Cesalpino by Ray and Linnaeus" (Morton 1981a: 129). But he also quotes, in a note, a letter by Cesalpino's contemporary Gaspar Bauhin, in which Bauhin states that Cesalpino's *De plantis* is "very obscure" and "doubt[s], wheter he would be intelligible to beginners and students" (quoted by Morton 1981a: 162).

119 Schmitt 1972.

120 Bellorini 2016: 68.

no's thought, Cesalpino did not follow blindly or slavishly Aristotle; rather, he reappropriates his thought, sometimes with a certain critical distance.

The introduction of philosophical considerations into a natural history treatise is very original. As well as its theoretical purport, it may also have had political implications in the context of the Renaissance universities. At that time, indeed, natural history observation was deemed "abasing" in contrast to "elevating" philosophical speculations.[121] By introducing Aristotelianism into botany, Cesalpino may have been attempting to render his project more theoretically respectable than the traditional medical treatises about plants, from which he wanted to separate his work. However, as a Renaissance scholar, Cesalpino sometimes questioned the limits of Aristotelianism, which he mastered perfectly, and proposed his own metaphysical ideas against the medieval tradition. He also challenged preconceived ideas or popular theses and was mindful of the scientific advances of his time. Cesalpino was accused of atheism and of pantheism for his theological and metaphysical conceptions, imbued with Averroism, although he always proclaimed his adhesion to the Christian faith.[122]

The relationship between science and philosophy is expressed in an ambivalent way in *De plantis*. Cesalpino certainly contributed to the development of modern botanical science, albeit from an Aristotelian perspective. For Aristotle, a science is a set of demonstrations based on universal principles, known by intellection, but obtained from sensations.[123] In the *Peripateticarum quaestionum libri quinque*, published in 1571, Cesalpino presented and defended his philosophical stance. The treatise opens with an exploration of the issue of method. According to Cesalpino, science must always go from what is best known to us, to what is best known in itself. Since sensation is the primary method for becoming familiar with an object, the first step, or starting point of scientific research must be inference from observable cases.[124] Indeed, individuals contain within them a singular form of the universal concept, that the soul of the observer

---

121 Bellorini 2016: 85.

122 His most virulent critics were Nicolaus Taurellus (1547–1606), author of *Alpes caesae* (1597), and theologian Samuel Parker (1640–1688), Bishop of Oxford, in *Disputationes de Deo et providentia divina* (1678).

123 Aristotle, *Analytica Posteriora* II, 19, 99b20–100b17 (ed. Ross 1949: 182–183). This passage is one of the most debated in the Aristotelian corpus. See in particular 99b20–22 and 32–36 (transl. Barnes 2002: 72–73): "you cannot understand anything through a demonstration unless you know the primitive immediate principles [...] We must therefore possess some sort of capacity – but not one which will be more valuable than these states in respect of exactness. And this is clearly true of all animals: they have a connate discriminatory capacity, which is called perception." – (Ὅτι μὲν οὖν οὐκ ἐνδέχεται ἐπίστασθαι δι' ἀποδείξεως μὴ γιγνώσκοντι τὰς πρώτας ἀρχὰς τὰς ἀμέσους, εἴρηται πρότερον. [...] ἀνάγκη ἄρα ἔχειν μέν τινα δύναμιν, μὴ τοιαύτην δ' ἔχειν ἣ ἔσται τούτων τιμιωτέρα κατ'ἀκρίβειαν. Φαίνεται δὲ τοῦτό γε πᾶσιν ὑπάρχον τοῖς ζῴοις. Ἔχει γὰρ δύναμιν σύμφυτον κριτικήν, ἣν καλοῦσιν αἴσθησιν·) See also Barnes' commentary on this question, pp. 259–271.

124 Cesalpino 1571: I, 1, 1C: *Sane horum causa est. Quoniam vnitatem prius cognoscimus quam multitudinem: vnum enim mensura est multitudinis, ipsa autem mensura in vnoquoque genere notissima est.*

will be able to identify from similarities between particular cases.[125] This knowledge, acquired from sensation, is erratic but indispensable.[126] The second step is the application of a method based on division, in order to discern the different intelligible realities at work in the similarities observed: thus the distinction between genera, species, parts, and concepts in general.[127] These do not necessarily correspond to commonly accepted notions, and the argument must therefore be thoroughly rewritten to align it with reality.[128] This paves the way for the third step, i.e. the definition of the various substances that have been identified.[129] Although they may seem contradictory, these methods are indeed three stages in a single process, which starts from the familiar chaos of sensation and moves towards the clarity of scientific intellection.[130]

Chapter 1 introduces general theoretical problems about the soul and heart of the plant. It replaces Aristotle's long developments dedicated to doxography and to the aporias linked to the object under study: Cesalpino relegates his – briefer – analyses of these issues to the last chapters of the first book. Thus, he remains faithful to the Aristotelian process for the foundation of sciences, even though he modifies the pattern usually followed by Aristotle in extant treatises. Chapters 2 to 11 correspond to the inference stage of the acquisition of knowledge about plants. In a less structured and systematic way, each chapter reproduces on its own scale the three stages of induction, division, and definition of the plant part studied in it. Chapter 12 proposes a method for dividing plant genera and subgenera, applied in chapters 13 and 14 to outline definitions.

The *Peripateticarum quaestionum libri quinque* is Cesalpino's most philosophical treatise. Each of its questions focuses on a topic which may be beyond Aristotelian

125 Cesalpino 1571: I, 1, 1C: *Cum igitur et totum et confusum vnitatem significet quae multitudinem seu distinctionem et partes continet, prius illa quam haec cognoscere oporter.*

126 Cesalpino 1571: I, 1, 1C: *Confusa enim est singularium cognitio, vniuersalium autem distincta.*

127 Cesalpino 1571: I, 1, 1C: *Hinc autem ex conceptibus maxime vniuersalibus, et indeterminationem multam habe, tibus, secundo loco per diuisionem ad species indiuiduas accedimus.*

128 For example, Cesalpino 1571: I, 1, 1C-D: *Vniuersalium enim ea prius mente concipimus, quae omnibus conueniunt, quam quae quibusdam vt ens et vnum prius, quam corpus, aut animal.*

129 Cesalpino 1571: I, 1, 1D: *Vltimo tendem species ipsas undiuiduas iterum exactius contemplantes in partes ex quibus earum substantia componitur, dissecamus. Atque hoc definitione praestatur.*

130 Cesalpino 1571: I, 1, 1D-E: *Triplici ergo progressu, vt in alii plerisque natura solet, perfectionem attingimus: inductione scilicet, diuisione, definitione. Inductione quidem similitudinem et conuenientiam intuemur: Diuisione, dissimilitudinem et differentiam: Definitione, propriam vniuscuiusque substantiam. Inductio ex singularibus vniuersalia facit, et materiam omnem intelligibilem menti offert: Diuisio differentiam vniuersalium inuenit tendens in ea quae specia sunt indiuidua: Definitio autem species in sua principia resoluit vsque ad alementa, incipiens a singularibus. Facilius enim est singular definire quam vniuersale, vt hominem quam animal: ideo a singularibus ad vniuersalia ascendendum esse praecipit Aristoteles 2.Post.tex.22.* Each stage of the argument relies or comments on passages from Aristotle's *Physica, Metaphysica, Analytica Posteriora* or *De Anima.* The link between demonstration and definition is established in question I, 2: any complete definition is, with a few nuances, a demonstration.

scientific knowledge. Cesalpino was proficient in zoology, anatomy, and familiar with recent advances in mechanics. For example, in his fourth book (question 5), he acknowledged the principle of the composition of motions, according to which a mixed motion follows the diagonal of the two motions producing it. Cesalpino illustrates this principle through the examples of a stone thrown forwards, and of raindrops on a tile, moving downwards under the effect of gravity while being deflected by the wind. "Contrary to a widely held belief among historians of science, the impossibility of the composition of a natural motion and a violent motion was not a dogma among all Aristotelians."[131] These mechanistic concepts, as well as those held by other contemporary professors in Pisa, such as Girolamo Borro (1512–1592), Francesco Buonamici (1533–1603), and Jacopo Mazzoni (1518–1598), most likely influenced Galileo's mechanics when he was a student, and later a professor of mathematics in Pisa. In any case, they provide a better understanding of an important aspect of Galilean science.[132] Similarly, Cesalpino's references to passages in his *Peripateticarum quaestionum libri quinque* are useful for the interpretation of some of the theses of *De plantis*.

Cesalpino's Aristotelian background emerges in his treatment of the concept of substance – *substantia*. This is the most common Latin translation of the Greek term οὐσία, proposed as early as Quintilian (1st century CE) and adopted in common usage from Marius Victorinus (290–364 CE) to Boethius (480–524 CE).[133] In Ancient Greek, including in Aristotle's writings, the term οὐσία (besides its generic meaning of "thing" or "wealth") can refer to the state of *being* as being, but also to what a given thing *is*; it can also simply designate *the thing* in question, or insist on its status as a *subject* that receives qualities or other determinations.[134] In other words, it combines the modern concepts of being, essence, material substance, and subject; it unifies or at least implicitly links them. Elsewhere, the Stagirite distinguishes only two meanings: individual subject and essential content.[135] This ambiguity resulted in competing translations, particularly in Latin Platonism, which favored the translation *essentia*. Authors such as Apuleius (125–170 CE) and Macrobius (370–430 CE) adopted this translation, probably following a suggestion made by Stoic philosopher Seneca (1–65 CE) but transferred the same ambiguities to this new term. Distinct Latin words (*substantia, essentia, materia, subiectum*...) only appeared with Augustine (354–430) and in medieval scholasticism, and became systematically employed to describe the different aspects of this conceptual field.[136]

131 Helbing 2008: 190. Although, he does not mention the historians by name.

132 Helbing 2008: 185–193. Galileo was, as a man of his time, an Aristotelian, but the very modern way in which he emancipated himself from Aristotelianism is much more complex than plain rejection.

133 Trego 2012: 233–256.

134 Aristotle, *Metaphysica* Z 1, 1028a10–31 (ed. Jaeger 1957: 128–129).

135 Aristotle, *Metaphyica* Δ 8, 1017b23–26 (ed. Jaeger 1957: 99–100).

136 Ghellinck 1942.

However, Cesalpino made no such distinctions in *De plantis*. He was not concerned with being as being, and did not seem to distinguish between a singular thing and that thing taken as the abstract support of its determinations. In his text, *substantia* refers to both the concrete matter of plants and their immaterial essence. He explains, for example, that the substance of bones is bounded by joints, that of roots is entangled, that of seeds is milky, that of flowers is light, while the substance of the pericarp corrupts after ripening.[137] He refers here to material substance, capable of physical transformation and spatial displacement. Moreover, "substance" is sometimes used interchangeably with "matter".[138] At the same time, he also defines the substance of plants as opposed to their accidents, according to the last difference between them, and to the purpose of the substance.[139] This substance is not material; it depends on the definition and classification of plants into genera and species. Such an investigation is a matter of form – the term "form" occasionally replaces "substance".[140] In Aristotle's works, form as an object of definition was already equivalent to substance understood as essential content; moreover, the substance of the living body is its soul.[141] Like Aristotle, Cesalpino purposefully used the same term to designate matter and form, but while Aristotle explicitly contrasted the two meanings, Cesalpino moved fluidly from one to the other. Paragraph 137 is exemplary in this respect: it defines the substance of the main genera (trees, shrubs, undershrubs, herbs) on the basis of the hardness of the substance or matter of the roots and shoots.

The entanglement of the two meanings is not without consequences. When Cesalpino wrote that teeth, hair, and horns are not really part of the substance of animals because they do not serve fundamental functions, or that the substance of seeds comes from the vital principle of the plant,[142] it is possible to understand his assertions in a physical as well as a metaphysical sense. This is not so much a case of unfortunate homonymy than the expression of an amphibological concept that cannot be subdivided into clearer concepts without altering its meaning. In the *Peripateticarum quaestionum libri quinque*, Cesalpino justified this association of matter and substance: he rejected the usual conception of matter as pure possibility, equating it instead to the body, possessing spatial dimensions, and, above all, a form, i.e., a substance.[143] All matter is organized from the outset, and therefore substantial, even if what makes it so is the form that structures it, which can therefore be defined separately as substance. This redefinition, of which we find an illustration in the first book of *De plantis*, is sig-

137 Respectively ch. 1: §2; ch. 2: §22; ch. 6: §61; ch. 7: §65 and ch. 9: §91.

138 For example, in ch. 6: §56.

139 Ch. 13: §135; §137; §143; ch. 14: §156.

140 Ch. 12: §127–129; ch. 13: §135–136.

141 Aristoteles, *Metaphysica* Z 17, 1041b25–32 (ed. Jaeger 1957: 165), and *De anima*, II, 1, 412a16–22 (ed. Ross 1961: 26), respectively.

142 Ch. 3: §27 and ch. 6: §56.

143 The genesis of this thesis in the thought of Simone Porzio is reconstructed by Colombero 1977: 280–283.

nificant. Indeed, Cesalpino also asserted that all substances are animate or compose an animate substance: there is therefore no inanimate reality that is substantial.[144] This implies that, if all matter is substance, there is no matter outside animate beings, since matter as indeterminacy awaiting substance is excluded.

Finalism, Nature, and God

According to Cesalpino, the classification of plants must rest upon the operations of nutrition and reproduction. The same applies to the description of plant parts, which occupies the first eleven chapters of *De plantis* Book I. Plant parts are analyzed above all according to their contribution to these operations, i.e., in a finalist manner. Defining the essence (the form or formal cause) from the end or purpose (the final cause) is not foreign to Aristotelian methodology,[145] and indeed finalist terminology is omnipresent in this first book. For instance, Cesalpino writes that the various husks are *dedicated to* (or, more literally: *given for*) the protection of the fruit, which exists for the *purpose of* reproduction:

> Qvemadmodvm in animalibus circumuolui foetum membranis oportuit ad tutelam, sic seminibus plantarum multiplicia *data sunt* inuoluctra ob eandem causam.
>
> Just as the fetus is suitably surrounded by membranes for its protection, so the seeds of plants *are endowed with* many husks for the same purpose.[146]

> Aliis autem natura praeter floris tegmen proprium inuolucrum seminibus *tradidit*.
>
> In others [plants], nature *gives* seeds a husk dedicated to protection in addition to the flower.[147]

> Quoniam folia *alterius finis gratia data sunt*, flores autem et florum tegmina solius *fructus gratia*.
>
> The leaves *are dedicated to another purpose*, whereas the flowers and the floral protections [are dedicated] uniquely *to* the protection of the fruit.[148]

144 Cesalpino 1571: I, 7.

145 See, for example, Aristoteles, *Metaphysica* Z 17, 1041a26–32 (ed. Jaeger 1957: 165); *Physica* II, 7, 198a24–27 (ed. Ross 1950: 198a). See Sauvé Meyer 1992; Henning 2009; Delcomminette 2018: 438–448.

146 Ch. 7: §64. Emphasis added.

147 Ch. 10: §97. Emphasis added.

148 Ch. 10: §107. Emphasis added.

Similarly, plant development takes place *in order* for growth and then reproduction *to* take place, enabled by daylight:

> Edendi fructus *gratia* germinationem *datam esse.*
>
> Development *is devoted to* the growth of the fruit.[149]

The adoption of this teleological model goes beyond the heuristic metaphor; the properties and very existence of the parts are justified by their function. The corruption of the parts is also explained by their purpose: leaves do not appear until they are required to protect the fruit, and they wither and fall when the fruit has grown large enough to need them no longer.[150]

The finalist framework is even clearer in Latin: wherever we have translated that plants are *endowed with* parts *dedicated to* a function, Cesalpino used the verb *dare*, "to give". The various parts are thus *given* to the plants for their protection, subsistence, or propagation. Yet, if a gift can be granted, it can also be refused. Thus, some plants receive a pericarp, while others, such as herbs or certain trees, are denied it. To compensate, these species receive a drier body that acts as a pericarp and fulfils the same protective function:

> Non omnibus autem datum est pericarpium ... quibus autem pericarpium negatum est, iis aliud corpus, quod eius vicem gerat, tributum est, siccius quidem pericarpio, sed non durius seminis cortice.
>
> Not all [plants] are endowed with a pericarp ... Plants that have been deprived of a pericarp have received another body that serves the same purpose, actually drier than a pericarp, but less hard than the bark of the seed.[151]

Cesalpino never describes a possible creator who may have caused or assigned these ends – the author of the gifts: God does not appear as such in *De plantis*. This absence is also characteristic of Cesalpino's scientific project about nature and of his explanation of living things.[152] However, Cesalpino was a theist, and in *De plantis*, Nature appears as a transcendent agent, which raises theological issues. The status of God is further discussed in the *Peripateticarum quaestionum libri quinque*: the divine Intellect must

149 Ch. 3: §29. Emphasis added.
150 Ch. 7: §67.
151 Ch. 10: §93.
152 It can be seen either as a modern innovation from the theological orientation of medieval conceptions of the natural world, or as a return to the old peripatetic methodology. However, this absence of reference to God leads one to relativize an interpretation of *De plantis* such as that of Atran (1990) who makes God a keystone of Cesalpino's conception of the species, of his classification, and of his supposed deviation from Aristotelian theses.

be pure self-contemplation, not creative activity, for it would then be distinct from the Supreme Good. But its status as a principle lies in its goodness, which makes it a universal final cause; the desire of material things *for* the final cause is responsible for the arrangement of those material things. The intervention of a demiurge is only relevant in a Platonic world, created from Ideas as the true principles, or in an artificial object resulting not from the natural powers of its components, but only from their juxtapositions. Cesalpino rejects both conceptions: since the order of the living is the actualization of such powers, it therefore lends itself to an organization by spontaneous desire without the need for a creator God to come out of his contemplation to create:

> Speculatiuus autem intellectus susbtantia existens per se, non alterius gratia appetibilis, ea quae apta nata sunt appetere, perficit, non agendo quippiam, aut introducendo in materia aliquid, hoc enim laboriosum esset; sed per sui praesentiam educendo ex potentia materiae formam ad quam apta nata est. Principium igitur motus horum in patiente solum, non in agente est, quae natura dicitur vniuscuiusque.
>
> Since the substance of the speculative Intellect exists *per se*, and is desirable on account of something which is not alien to it, this substance brings to perfection that which is by nature able to desire, not by acting in any way, nor by introducing anything in the matter, for this would be laborious. On the contrary, it leads by its own presence, out of the potency of matter, the form that it is naturally able to receive. Therefore, the principle of its motion lies only in the receiver, not in the agent, and that is what is called nature in everything.[153]

Once again, Cesalpino's appears as an orthodox Aristotelian, embracing the concept of God as the first unmoved mover developed in the *Metaphysics*.[154]

A general tendency of Renaissance philosophy was to preserve teleological accounts of nature found in earlier thinkers, but to propose a different interpretation for them. While several medieval authorities considered final causes as corresponding to the will of the Christian, personal God, most Renaissance thinkers attributed these ends to nature itself.[155] So did Cesalpino, who repeatedly mentioned nature as an agent who provides plants with what they need, divides them strategically so that their functions are best fulfilled, and distributes its resources sparingly but beautifully.[156]

> Oportuit calorem innatum foueri in seminibus, ne exterius transpirando, ipsa euanida redderentur. Huius igitur gratia natura semina obduxit alio corpore, vt plurimum carnoso et humecto: que-

---

153 Cesalpino 1571: II, 4: 29c-d.

154 Aristotle, *Metaphysica* Λ 9–10, 1074b15–1075a25 (ed. Jaeger 1957: 258–260).

155 Ingegno 1988; Fornasier 2020. Such a shift in philosophical conceptions is by no means consensual: for example, philosopher Philip Melanchthon (1497–1560) adhered to the medieval conception of a transcendence of natural teleology against any effort to make it more immanent in his *Initia doctrinae physicae*: Melanchthon 1549 (ed. Brettschneider 1846: 2.307–308 and 349–354 in the *Corpus reformatorum* 13).

156 Ch. 4: §41; ch. 5: §50; ch. 7: §74; ch. 8: §78–79; ch. 9: §86; ch. 10: §96–97; ch. 11: §114; ch. 13: §142–143 and ch. 14: §150. See also ch. 6: §5 and ch. 7: §69.

> madmodum Medici tuberculis concoquendis cathaplasma aliquod superponere solent, vt humiditas cum calore in parte conseruetur.
>
> The innate heat has had to be preserved in the seeds, so that it does not disperse to the outside and weaken them. That is why nature covers the seeds with another body (fleshy and moist in most cases), in the same way that physicians are accustomed to applying poultices to warm tumors, so that the moisture is preserved with the heat in this part.[157]

Study of *De plantis* alone is insufficient to determine whether this understanding of nature as responsible for the properties of plants is to be taken literally, or if it is merely a way of indicating their finalized orientation. This ambiguity may have contributed to Cesalpino being – wrongly – accused of pantheism. The term "nature" is just as often used to designate the nature of plants or of certain parts of them, such as the bark or pith, without any reference to a demiurgic agent. Comparison with the *Peripateticarum quaestionum libri quinque* sheds light on this question: nature is not a demiurge; it manifests itself in the spontaneous tendencies of the powers of living bodies. These tendencies are finalized, oriented towards the goodness of the God-Intellect, who acts more as a final than an efficient cause upon their desires and, thus, upon the being that they actualize. It is therefore the nature of living beings that makes them what they are, in a finalized way, and there is no need for this nature to be an independent reality.[158]

Purpose is not only operating at the level of Nature as a whole, but also within each plant. The starting point of *De plantis* is Aristotle's psychology: at least in his botany, Cesalpino used concepts, problems, and arguments from Aristotle's *De anima* much more often and consistently than from any other source. Therefore, we shall examine what he borrowed from this text to structure his own research. Aristotle concluded the first book of his treatise with the observation that plants and animals have at least some life principle in common, which allows survival without sensation, and which he calls a soul:

> Ἔοικε δὲ καὶ ἡ ἐν τοῖς φυτοῖς ἀρχὴ ψυχή τις εἶναι· μόνης γὰρ ταύτης κοινωνεῖ καὶ ζῷα καὶ φυτά, καὶ αὕτη μὲν χωρίζεται τῆς αἰσθητικῆς ἀρχῆς, αἴσθησιν δ' οὐθὲν ἄνευ ταύτης ἔχει.
>
> It also seems that the first principle in plants is a sort of soul; for animals and plants have this alone in common, and this exists separated from this perceptive first principle, though nothing lacking this has perception.[159]

157 Ch. 9: §86.

158 In the 17th and 18th centuries, Cesalpino was accused of denying divine providence, either through Aristotelianism or through a form of proto-spinozism. See on this subject Colombero 1980. On this point at least, the first accusation is not unfounded.

159 Aristotle, *De anima*, I, 5, 411b27–30 (ed. Ross 1961: 25; Engl. transl. Shields 2016: 21).

Plants are alive because they are able to assimilate food and to grow, although they do not possess the sensations, desires, and faculty to move, characteristic of animals: this nutritive faculty is common to all living things.[160]

> Ὑπάρχει δὲ τοῖς μὲν φυτοῖς τὸ θρεπτικὸν μόνον, ἑτέροις δὲ τοῦτό τε καὶ τὸ αἰσθητικόν. Εἰ δὲ τὸ αἰσθητικόν, καὶ τὸ ὀρεκτικόν· ὄρεξις μὲν γὰρ ἐπιθυμία καὶ θυμὸς καὶ βούλησις.
>
> The nutritive faculty alone belongs to plants; both this and the perceptual faculty belongs to others. But if the perceptual faculty, then also the desiderative faculty: desire is appetite, spirit, and wish.[161]

The principal operations of the *nutritive* (elsewhere also called *vegetative*) faculty or soul are twofold: nutrition itself, and generation. The former includes obtaining and processing food; the latter designates the identical reproduction through which a species survives:[162]

> [...] φυσικώτατον γὰρ τῶν ἔργων τοῖς ζῶσιν, ὅσα τέλεια καὶ μὴ πηρώματα ἢ τὴν γένεσιν αὐτομάτην ἔχει, τὸ ποιῆσαι ἕτερον οἷον αὐτό, ζῷον μὲν ζῷον, φυτὸν δὲ φυτόν, ἵνα τοῦ ἀεὶ καὶ τοῦ θείου μετέχωσιν ᾗ δύνανται· [...] Ἐπεὶ οὖν κοινωνεῖν ἀδυνατεῖ τοῦ ἀεὶ καὶ τοῦ θείου τῇ συνεχείᾳ, διὰ τὸ μηδὲν ἐνδέχεσθαι τῶν φθαρτῶν ταὐτὸ καὶ ἓν ἀριθμῷ διαμένειν, ᾗ δύναται μετέχειν ἕκαστον, κοινωνεῖ ταύτῃ, τὸ μὲν μᾶλλον τὸ δ’ ἧττον, καὶ διαμένει οὐκ αὐτὸ ἀλλ’ οἷον αὐτό, ἀριθμῷ μὲν οὐχ ἕν, εἴδει δ’ ἕν.
>
> For the most natural among the functions belonging to living things, at least those which are complete and neither deformed nor spontaneously generated, is this: to make another such as itself, an animal an animal and a plant a plant, so that it may, insofar as it is able, partake of the everlasting and the divine [...] Since, then, these things are incapable of sharing in the everlasting and the divine by existing continuously (because among perishable things nothing can remain the same and one in number), each has a share insofar as it is able to partake in this, some more and some less, and remains not itself but such as it is, not one in number but one in form.[163]

160 Aristotle, *De anima*, II, 2, 413a25–b9 and II, 3, 414a29–b6 (ed. Ross 1961: 28–31).
161 Aristotle, *De anima*, II, 3, 414a29–b6 (ed. Ross 1961: 31; Engl. transl. Shields 2016: 27).
162 Aristotle, *De anima*, II, 4, 415a23–b7 and 416a19–b31 (ed. Ross 1961: 33–37). On the history of the vegetative soul, see Mix 2018; Baldassarri and Blank 2021; for the explanation of the concept in Aristotelianism, see specifically: 43–53, 55–66, 117–128.
163 Aristotle, *De anima*, II.4, 415a27–b1 & b3–7 (ed. Ross 1961: 34; Engl. transl. Shields 2016: 29). Atran (1990: 138–142, 152–153) argues that Cesalpino's concept of species is more absolute than Aristotle's, due to Christian (Neoplatonic) influences. It would have led Cesalpino to consider natural species as eternally fixed, and individual properties as necessarily defined by their essence in relation to "God's eternal plan" (Atran 1990: 142). Atran even states that "Cesalpino was for all intents and purposes the originator of the taxonomic concept of the eternal species" (1990: 227). However, the references to *De plantis* Book I mentioned by Atran (1990: 27; 140; 141; 142) are often given as page numbers rather than as textual quotations, and do not always support the idea that Cesalpino's concept of species is wholly different from Aristotle's, especially when the full translation of Book I (below), some passages from Aristotle such as the one above, and the fact that Cesalpino does not mention God anywhere in *De plantis*, are taken into account.

These functions structure the first chapter of *De plantis*, and the sequence of the following chapters within the methodological moments: the division of plants into fundamental parts, then botanical classification itself. The first essential operation of the plant is nutrition, which is effected by the root; as such, the root acquires pre-eminence and becomes a suitable criterium for the definition of a first level of subgenera:

> Est autem primum uegetatiui opus, quod omnibus viuentibus inest alimenti attractio, quo nutriantur et crescant.
>
> The first operation of the vegetative [soul] – namely, as in all living things, the acquisition of food – is that which allows it to feed and grow.[164]

Growth is a complementary aspect of this nutritional operation. The second essential operation is reproduction:

> Secundum autem vegetatiui opus est generare sibi simile, quod et perfectione prius est, cuius gratia dati sunt fructus et partes ad fructificationem facientes.
>
> Furthermore, the second task of the vegetative [soul], which is first as far as perfection is concerned, is to reproduce identically, for which the fruit and the reproductive parts of the plant are designed.[165]

This is carried out by the upper part of the plant, which Cesalpino calls either a stem or a shoot, the criterium for the definition of a second level of subgenera:

> Quoniam autem altricis animae opus est gignere quale ipsum, siue id ex alimento fiat ad conseruationem singulorum, siue ex semine ad specierum aeternitatem: duae ad summum partes perfectioribus datae sunt, maxime necessariae: Vna, per quam sumant alimentum, quae radix appellatur. Altera qua fructum tanquam foetum ad speciem propagandam ferant, qui caulis vocatur in humili materia, caudex autem in genere arboreo.
>
> But since an operation of the nutritive soul is to reproduce identically (which it does either through food for the preservation of the individual, or through the seed for the continuation of the species), plants must necessarily be endowed with two main parts to be complete. One of these, through which they procure food, is called the root; the other, through which they bear both fruit and progeny in order to propagate the species, is called a stem in small plants, and trunk in the genus of trees.[166]

Together, these fundamental parts determine the four main plant genera and the two taxonomic levels of division (subgenera), below which there are only species.[167]

164 Ch. 13: §137–139.
165 Ch. 13: §140.
166 Ch. 1: §5.
167 These two levels correspond roughly to a family taxon and a genus taxon in their current sense.

The Analogy

Although Cesalpino conceived finality as a real orientation of nature rather than a mere image, he often used analogy to describe and explain plant functioning.[168] The most common is the analogy with animals: plant parts and processes are mapped onto similar features observed in animals.[169] This practice follows Aristotelian methodology: less familiar concepts are explained through reference to more familiar ones. Thus, plants are – at first – considered through the lens of zoological concepts. In animals, functions appear directly linked to body parts and organs (for example, the mouth allows to eat, the stomach to digest, the legs to move, the eyes to see), which is not the case in plants. Plant parts are less differentiated, and functions less obviously and less clearly linked to organs.

The first chapter of *De plantis* opens with a comparison – a contrast, even. As noted above, Cesalpino's starting point is the Aristotelian doctrine of the soul, divided into parts or faculties. Animals have sensation, desire, and locomotion (and humans also have reason), as well as nutrition and reproduction; plants only possess the last two faculties, because of their strictly vegetative soul. In the first paragraph, Cesalpino, following this doctrine, states that plants are fundamentally distinct from animals because they have no sensory nor locomotor faculty, not by accident, but because their soul, that is to say, their substance, is different. Any comparison between plants and animals is therefore to be understood *mutatis mutandis*: it rests upon structural or functional analogies rather than on real similarities (hinting at, for example, a common origin). Nevertheless, the correlation is warranted, since plants and animals have in common the functions of nutrition and reproduction, managed by the vegetative soul: on these points at least, the analogy is more than a metaphor. Cesalpino thus justified and qualified his application of animal categories even when observable resemblances are scarce.

The analogy highlights the similarity of the function performed by parts that seem to have little in common: the veins of animals and the roots of plants. According to Cesalpino, both allow food to be transported from an organ of nutrition to a principle that extracts the vital heat. Thus, the earth plays the same role for plants as the belly for animals, because the roots are connected to it like veins to the entrails:

> Natura enim venarum, quae alimentum ex ventre hauriunt, vt illud in vniuersum corpus distribuant; aliqua ex parte respondere videtur cum plantarum radicibus; nam similiter hae ex terra tanquam ex ventre cui implantantur trahunt alimentum.

> However, the nature of veins, which bring food from the abdomen and distribute it throughout the

168 On the role of analogy in Cesalpino's botany, see Hiernaux and Tresnie 2023b.
169 On animal analogy in *De plantis*, see Atran 1990: 224–230.

> body, seems to partly correspond to the roots of plants. Indeed, the two embed themselves in the earth and in the abdomen in a similar way and draw food from them.[170]

This analogy is already present in Aristotle:

> Τὰ μὲν γὰρ φυτὰ λαμβάνει τὴν τροφὴν κατειργασμένην ἐκ τῆς γῆς ταῖς ῥίζαις (διὸ καὶ περίττωμα οὐ γίνεται τοῖς φυτοῖς· τῇ γὰρ γῇ καὶ τῇ ἐν αὐτῇ θερμότητι χρῆται ὥσπερ κοιλίᾳ).
>
> Plants get their food from the earth by their roots; and since it is already treated and prepared no residue is produced by plants – they use the earth and the heat in it instead of a stomach.[171]

> Τὰ μὲν οὖν φυτὰ τὰς ῥίζας ἔχει εἰς τὴν γῆν (ἐκεῖθεν γὰρ λαμβάνει τὴν τροφήν), τοῖς δὲ ζῴοις ἡ κοιλία καὶ ἡ τῶν ἐντέρων δύναμις γῆ ἐστιν, ἐξ ἧς δεῖ λαμβάνειν τὴν τροφήν· διόπερ ἡ τοῦ μεσεντερίου φύσις ἐστίν, οἷον ῥίζας ἔχουσα τὰς δι' αὐτῆς φλέβας.
>
> The roots of plants are of course in the ground, because that is the source from which plants get their nutriment. For an animal, the stomach and the intestines correspond to the ground, the place from which the nutriment has to be derived. And the Mesentery exists to contain these vessels, corresponding to roots; they pass through the inside of it.[172]

Cesalpino qualifies this parallel further in chapter 2;[173] here, he clarifies the concept of plant nutrition by analogy to animal nutrition, which depends on the same type of nutritive soul. A similar comparison can be drawn between budding and embryogenesis, although this time, embryology benefits from the botanical model; the development of the foetus, which cannot be observed, becomes clearer if we relate it to that of the shoot:

> Foetus enim, qui adnascitur, tamquam germen eius partis viuit nutrimento ex eadem affluente. Differt autem; quoniam in his principium extrinsecus ducitur, scilicet ex semine maris, alimentum ex vtero trahente; in plantis autem tum materia, tum principium intrinsecus prouenit.
>
> The fetus, in fact, which "sprouts" like a sort of shoot in its own way, survives thanks to the food that is provided, in the same way that a shoot does. There is a difference, however, in that in animals, the principle [of development] comes from the outside, that is by the semen of the male, although the food comes from the uterus. In plants, on the other hand, both the matter and the principle come from within.[174]

170 Ch. 1: §3.
171 Aristotle, *De partibus animalium*, II, 3 650a20–23 (ed. and Fr. transl. Louis 1956: 32; ed. and Engl. transl. Peck 1945: 135)
172 Aristotle, *De partibus animalium*, IV, 4, 678a3–16 (ed. and Fr. transl. Louis 1956: 111; ed. and Engl. transl. Peck 1945: 313–315).
173 Ch. 2: §18.
174 Ch. 3: §28.

Cesalpino carried the analogy further: since veins and roots have the same nutritive function, they must be connected to parts that are also similar in animals and plants. This is Cesalpino's solution to the old Aristotelian problem of the identification of the seat of the life principle in plants. Indeed, to dispense life, this principle must be located in a strategic place, from which it can coordinate the distribution of heat and vital fluid extracted from absorbed nutrients. The soul is of course the entelechy of the living body in its totality, but the specialization of different parts in living beings suggests that the soul does not reside uniformly in all parts. For Aristotelians, the case of animals is easy to explain, since the heart is an ideal candidate for the location of the irreplaceable seat from which vital heat is distributed to the organs. Analogical reasoning allowed Cesalpino to assert the existence of a plant heart, at the point where the root and stem meet, i.e., the root collar:

> Duae autem sunt partes plantarum maxime conspicuae radix scilicet, et id totum, quod sursum attollitur, merito in intermedio, qua scilicet radix germini coniungitur, locus videatur cordi plantarum oportunissimus.
>
> To sum up, there are two distinct parts in plants, namely the root and everything that grows from bottom to top. The place which connects the root to the shoot [the root collar] seems to be a very appropriate location to find the heart of plants.[175]

Far from being an organ as complex as the animal heart, which is very different in appearance, this plant heart takes the form of the pith of the stem that originates there (just as animal marrow spreads along the spine from the head).[176] This idea had already been briefly proposed by Aristotle:

> Πᾶσι γὰρ τοῖς ζῴοις τοῖς τελείοις δύο τὰ ἀναγκαιότατα μόριά ἐστιν, ᾗ τε δέχονται τὴν τροφὴν καὶ ᾗ τὸ περίττωμα ἀφήσουσιν· οὔτε γὰρ εἶναι οὔτε αὐξάνεσθαι ἐνδέχεται ἄνευ τροφῆς. [...] Τρίτον δὲ μέρος ἐν πᾶσίν ἐστι τὸ τούτων μέσον, ἐν ᾧ ἡ ἀρχή ἐστιν ἡ τῆς ζωῆς.
>
> [I]n all living creatures of perfect formation there are two parts most necessary above all: one by which food is taken in and the other by which residues are eliminated [...] And in all creatures there is a third part intermediate between these indispensable two, and this is the seat of the source and principle of life.[177]

175 Ch. 1: §13.

176 See ch. 1: §8–14.

177 Aristotle, *De partibus animalium*, II, 10, 655b30–33 and 37–38; cf. II, 3, 650a3–7; and II, 10, 655b29–656a7 (ed. and Fr. transl. Louis 1956: 48, 48, 31–32, and 48, respectively; ed. and Engl. transl. Peck 1945: 171–173, 173, 133, and 171–173, respectively). See also *De respiratione*, XVII, 4, 478b31–479a1 (ed. Ross 1955: 478–479).

Animal analogy thus provides a more efficient basis than observation for the discussion of the existence and function of the plant heart; these issues will be further qualified through arguments that we shall analyse below.

Adhesion to Aristotle's doctrine includes sharing his aporias. Nutrition, common to both plants and animals, involves a certain selection of food. In animals, it takes place through sensation, via the sensitive soul; the analogy ceases to be relevant, since plants do not possess a sensitive soul. Therefore, Cesalpino endeavored to replace the metaphysical explanation of selective nutrition in animals with a material explanation in plants. He favored models belonging to the inanimate, even artificial world: the force of attraction of a magnet, the vacuum of a cupping-glass, the absorption of a sponge, and finally capillary filtration in oil lamps. He selects the latter as the most adequate:[178]

> An quaedam sicca secundum naturam humorem trahunt? vt lintea, spongiae, pulueres: quaedam respuunt, vt auium quarundam plumae, Adiantum haerba, haec enim et si in aqua degant, non tamen madefiunt aqua ipsis non haerente: illa autem multum sorbent; quia cum aqua conueniunt magis, quam cum aere. Ex huiusmodi igitur natura partes illas in plantis constare putandum est, quibus vtitur anima altrix ad trahendum alimentum: Idcirco eae non ad venarum similitudinem meatu quodam continuo peruiae sunt: sed potius instar neruorum ex villosa constant substantia: sic enim bibula earum natura continue humorem ad principium caloris innati ducit, vt in lucernarum luminibus videmus; funiculo enim quodam vtuntur, quo oleum continue ad flammam ducatur.
>
> Might this be a case of dry things which by nature attract moisture? This is the case of linen, sponges, and dust. But other things repel moisture, like the feathers of certain birds or the fronds of southern maidenhair fern: even when you soak them in water, they are not dampened, because the water does not attach to them. [Among these things,] the former absorb a lot of water, because they are more compatible with water than with air. We must therefore consider that the parts of plants that the nutritive soul uses to obtain food are of a similar nature. As a result, the roots do not travel through the plant in a continuous path as veins do, but just like nerves, consist of a tangled substance. Thus, their absorbent nature drives the liquid in a continuous way towards what produces the internal heat, as we can see in the lights of lamps, which use a wick through which oil continuously supplies the flame.[179]

Aristotelian method and psychology led to the description of plants by analogy with animals. Because of the limitations of this method, Cesalpino introduced in the descriptions analogies to the physical world and from crafted objects: this is a first step towards a mechanistic conception of plants.

Thus, if the animal analogy is philosophically sound, it is not always sufficient to describe plants. Other metaphors were convenient and fruitful, but theoretically more fragile, and Cesalpino adapted his vocabulary accordingly: it is, "so to speak" (*quasi*), of their own accord, spontaneously, that the roots cause divisions in the plant nodes; the

178 Ch. 2: §19–23.
179 Ch. 2: §21–22.

heart is "like" (*tanquam*) a workshop; the substance of flowers is "a little like a breath" (*aliquem spiritum*); the tendrils of climbing plants are "like" (*tanquam*) hands.[180] Words such as *quasi*, *veluti* or *ut*, or their combination, introduce cautious reservations about the suggested metaphors. At times, such terms modify the animal analogy itself: the stem is only a "sort of" (*quasi*) spine along which the marrow spreads, from the root which is only "like" (*tamquam*) a head:

> Pulcherrime autem huic parti conuenire videtur cerebri nomen, quemadmodum enim in animalibus cerebri medulla in capite est, vnde spinalis medulla exoritur in totam spinae longitudinem diducta, sic in plantis cerebrum in radice *tamquam in capite* sedens per totum caulem *quasi per spinam* dorsi medullam deducit ad vitalem humorem ramis et extremis surculis distribuendum.
>
> The name "brain" seems most elegantly suited to this part. After all, in animals, brain marrow is found in the head, from which the marrow of the spinal cord originates and stretches out along its whole length. Similarly, in plants, the "brain" is situated in the root *as in a kind of head*, which spreads the pith up throughout the length of the stem, *a little like along a backbone*, in order to distribute the vital liquid all the way to the furthest twigs and scions.[181]

The fact that leaves and flowers exist solely for their function is sometimes also qualified in this manner.[182] These terms may reflect Cesalpino's awareness that his method leads to purely functional finalism; in the absence of such reservations, nature would seem to organize plants solely for growth and reproduction rather than for beauty or harmony in the world.

Positioning of Cesalpino in the History of Classification

During the 16th century, the number of plants discovered in, and imported from, other continents increased significantly. In consequence, Cesalpino believed that botanists and physicians should order this overwhelming amount of knowledge in a more efficient and scientific way than Renaissance common sense did.[183] Following in the footsteps of his master Luca Ghini, Cesalpino was the first scholar to produce a stable, coherent classification of plants resulting from "serious philosophical reflection."[184]

---

180 Ch. 2: §15; ch. 4: §40; ch. 7: §65; ch. 11: §114.
181 Ch. 1: §14. Emphasis added.
182 Ch. 3: §30; Ch. 7: §67.
183 Ogilvie 2006: 225–226. Mayr (1982: 58) notes that before the 16th century, since "relatively few plants were known, one could find a species by simply thumbing through an herbal until one encountered something reasonably similar."
184 Bellorini 2016: 77.

His principles of classification were scientific inasmuch as they relied on the observation of morphological similarities and differences. They also allowed him to develop a predictive method for determining how to classify yet undiscovered plants species, and even genera.[185] Although botanists such as Mathias de L'Obel had previously attempted to order botanical species, Cesalpino was the first one to analyze, define, and justify his criteria for classification.[186] All the subsequent classifications "up to and including Linnaeus were variations and improvement upon the approach first taken by Cesalpino."[187]

De plantis was one of the first, if not the first, treatises on botany in the modern sense, insofar as it emancipated itself from pharmacotherapeutics, agronomy, or other practical botanical knowledge in its theoretical undertaking. The challenge was to understand and describe plants for their own sake, and not for their uses.[188] When Cesalpino explained that medicinal properties were not relevant for botanical classification, he explicitly distanced himself from the medical approach of Dioscorides, who ordered plants according to the usefulness of their parts from which juices, gums, and other products were extracted:

> At Dioscorides tamquam Medicus solum communionem circa facultates Medicas accepit, quo ordine succos, lachrymas, radices, semina, & alias plantarum partes persecutus est.
>
> But Dioscorides, as a physician, only accepted the classification according to medical properties, on the account of which the saps, the tears, the roots, the seeds, and the other parts of plants are ordinarily looked for.[189]

Cesalpino did not lack interest in the medicinal properties of plants but considered them inessential. As he remarked in the dedication letter prefacing his treatise:

> Facultates quoque addere superfluum duxi: cum enim hae apud multos authores praecipue Dioscoridem et Galenum admodum copiose legantur.
>
> I have considered it superfluous to add the medical properties; in fact, these have been expounded by many authors, and above all, at great length by Dioscorides and Galen.[190]

185 Jensen 2001; Bellorini 2016: 79.

186 Ogilvie 2006: 221; Bellorini 2016: 81. L'Obel's classification (1571) was also based on morphology, but arbitrarily built on the sole criterion of the leaf form, which resulted in artificially grouping together very different plants. On de L'Obel's botany, see Arber 1912: 78–79; Greene 1983: 2.847–876; Reeds 1991: 66–72; Ogilvie 2006: 44–46; Magnin-Gonze 2009: 80–81.

187 Mayr 1982: 161.

188 Jensen 2001; Ventura 2013.

189 Cesalpino 1583: Dedication letter V.

190 Cesalpino 1583: Dedication letter VI, translation adapted from Bellorini 2016: 77.

According to Morton, he mentioned therapeutic properties only when they were previously unknown or for the purpose of botanical discrimination; otherwise, he referred to medical writers such as Disocorides and Galen.[191] In his *Outlines of the history of botany*, Scottish botanist Robert John Harvey-Gibson (1860–1929) even declared that *De plantis* "is at last divorced from materia medica."[192] This emancipation from practical concerns would paradoxically proved beneficial to medicine. By studying the *differentiae*, or natural affinities of plants for the purpose of systematic classification, *De plantis* not only avoided misidentifying plants, and thus also medicinal plants, but also suggested common therapeutic properties among (new) species grouped together in a more meaningful manner.[193]

Cesalpino did not present his work as particularly polemical in this respect, although he asserted and justified his position on several occasions. The clearest cases are the two instances in which he explicitly opposes the ways of "herbalists" and "physicians" to his new plant classification:

> Herbae autem proprie appellantur, quae postquam semen perfecerunt, moriuntur ex toto, vt Fruges et Olera plurima, quamuis alio modo Herbarij accipiant, herbas appellantes, quae in foliis insitas habent vires, vt etiam rustici appellare solent, quarum folia ad iumentorum pabulum veniunt, et ex quibus foenum parant.
>
> Strictly speaking, the term "herbs" is used for those plants which, after having produced their seed, die completely, as is the case with many vegetable varieties and legumes. That said, herbalists understand the term differently. They use "herbs" for those plants whose medicinal properties reside in the leaves, as they also usually qualify with "rustic" those whose leaves, with which one prepares hay, are used to feed draught or pack animals.

> Quaedam tamen sequuntur naturam specificam, vt plantarum vires in medicinis, sapores, et aliae proprietates, quas Medici in primis considerant, quae etsi aliquando per se insint, non tamen differentiae sunt constituentes earum substantiam, quas hoc loco quaerimus.
>
> Finally, the differences that we are looking at here and which are a result of specific nature, like the medicinal properties of plants, their taste and other attributes in which physicians are primarily interested, are however not constituent to the substance, even if they are somehow present *per se*.[194]

Cesalpino, a physician himself, presents his argument here as a botanist, an interesting dissociation of roles in the 16th century. However, the use of the first person plural

191 Morton 1981a: 141. This innovative way of studying botany independently of medicine influenced later scholars such as Adam Zalužanský (ca. 1555–1613) (Harvey-Gibson 1919: 19–20; Morton 1981a: 162).
192 Harvey-Gibson 1919: 19.
193 Cesalpino 1583: Dedication letter IV–V. See also Ogilvie 2006: 223.
194 Ch. 12: §122–123 and ch. 14: §156.

often refers to a generic subject, and more rarely to Cesalpino himself as the author (of *De plantis* or the *Peripateticarum quaestionum libri quinque* when he referred to them as *Quaestiones peripateticae*). From the chapters on classification itself and beyond,[195] "we" designates Cesalpino more systematically as a botanist and researcher, sometimes challenging other existing classifications. On occasion, Cesalpino quotes the argument of an anonymous scholar, for example:

> *Illud argumento est, quod...*
>
> One argument follows that...[196]

He often attributes a thesis to a subject in the third person plural, but this may also indicate a commonly accepted truth that he does not seek to refute; the counterfactual mode is the clearest mark of Cesalpino's rejection of a proposition, preceding the mention of an absurd, undesirable, or empirically false consequence of such a proposition.[197] It is, however, difficult to determine whether these propositions have really been submitted by Cesalpino's colleagues, or whether they are merely theoretical possibilities that are worth considering, but must be discarded. These textual observations, combined to the dedication letter prefacing *De plantis*, betray Cesalpino's conscious assertion of a new, personal, and original method of classification, in contrast to those adopted by the Ancients and by his contemporaries.

Linnaeus considered Cesalpino to be "the first orthodox systematist,"[198] and the Italian botanist is still often acknowledged as the initiator of scientific classification.[199] Indeed, *De plantis* does not simply list plants according to their names or uses, but classifies them according to their morphology and to the affinities between the genera derived from them. Cesalpino explicitly rejects alphabetical classification systems:

> Aliis, quo facilius memoriae mandarentur, placuit eo ordine digerere, qui secundum nominis incipientes literas datus est: sed hic tamquam maxime fallax, et longissime a rei natura discedens a grauioribus huius scientiae authoribus reprobatus est.
>
> Others deemed good, in order to remember them more easily, to divide them in an order that is given following the first letters of the name; but this was rejected, by the most serious authorities of this science, as the most false, so to say, and departing the furthest away from the nature of reality.[200]

195 From ch. 11: §117 specifically.
196 Ch. 3: §32.
197 Ch. 12: §131–133; ch. 13: §136; §138; §142–143; ch. 14: §147.
198 Stafleu and Cowan 1976: 1.478.
199 Morton 1981a: 137; Greene 1983: 2.808; Buoncristiani 2003; Bellorini 2016: 68.
200 Cesalpino 1583: Dedication letter V. On this passage, see Heideklang 2023.

Famous naturalist Jean-Baptiste de Lamarck (1744–1829) wrote accordingly in the preliminary speech of his *Encyclopédie méthodique*:

> Cesalpino [...] must therefore be regarded as the first botanist who tried to find a method by which plants could be most reliably recognized or determined.[201]

Cesalpino observed anatomical details and discovered the basis of phyllotaxy.[202] While he adopted the very general morphological classification of plants into trees, shrubs, undershrubs, and herbs inherited from Theophrastus, he introduced truly innovative taxonomic approximations based on more precise natural affinities. Cesalpino considered the structure of flowers, fruits, and seeds as the key to these affinities, which should be reflected in a classification of plants.[203] However banal this idea may seem today, Greene emphasizes its remarkable, historical significance. *De plantis* is:

> the earliest book in which all genera of plants are arranged in a sequence thought to represent their natural affinities, the work in which the proposition that by their fruits the subjects of the vegetable may be philosophically and scientifically coordinated was first carried into effect, remembering that upon this principle as fundamental, all systematic botany has been developed.[204]

And yet, although natural classification constituted the principle of modern systematics, the publication of *De plantis* alone was not enough to supplant the more artificial methods of Cesalpino's successors:

> When Cesalpino issued his great book of taxonomic revolution, botanists hardly knew what it meant, and about two generations more of them let the volume moulder on the book shelves unregarded.[205]

Sachs remarks that it was not until some thirty years after its publication that the innovations of *De plantis* were identified in Gaspard Bauhin's work.[206] For Atran, the

201 "Aussi Cesalpino [...] doit-il être véritablement regardé comme le premier botaniste qui essaya de trouver une méthode au moyen de laquelle les plantes seraient les plus sûrement reconnues ou déterminées" (Lamarck 1783, quoted by Magnin-Gonze 2009: 76).
202 Condorelli 1975.
203 Sachs 1875 (Engl. transl. Garnsey 1890: 5–12, 37–58); Sprague 1950: 1–23; Bremekamp 1952; Callot 1971: 173–177; Larson 1971: 20–38. Cesalpino did not neglected other secondary morphological properties, even if they stayed implicit and are mainly visible in his classificatory decision themselves: see Morton 1981a: 160–161.
204 Greene 1983: 2.809.
205 Greene 1983: 2.637.
206 Sachs 1875 (Engl. transl. Garnsey 1890: 38); Singer 1950: 174. On Bauhin, see among others: Arber 1912: 93–98, 130–132; Reeds 1991: 110–133; Ogilvie 2006: 210–219; Magnin-Gonze 2009: 82–87.

real influence of Cesalpino on later systematic botany manifested itself in the work of Joachim Jung (combined to Galileo's influence).[207]

According to Sachs, Cesalpino's theoretical desire to create a natural method of classification based on observed affinities between species led to a classification system that was partly artificial, insofar as its criteria were set *a priori*.[208] Historian of botany Harvey-Gibson's assessment is even more pejorative; he considers that Cesalpino built his classification with "Aristotelian obsessions"[209] and "sophistries"[210] by "meditating on theoretical principles" establishing "groups in the highest degree unnatural, embracing plants which [...] had not the slightest relationship to each other."[211] Agnes Arber (1879–1960), researcher in plant morphology and in history of botany at Cambridge,[212] also rates Cesalpino as a minor botanist, for similar reasons:

> His general tendency seems to have been to narrow the field of systematic criteria, and this was a retrograde step. The empirical development of taxonomics at the hands of the Renaissance herbalists gave surer guidance than Cesalpino's *a priori* views.[213]

Although Sachs and Arber's were important reference works in 20th-century history of science,[214] this summary trial of Cesalpino's systematic work is too hasty;[215] indeed, the relationship between empirical observation and the use of "*a priori* views" is more complex and subtle than it appears.[216] Moreover, Linnaeus himself, in the 18th century, distinguished himself by publishing his artificial system of classification according to sexual characteristics; its most accomplished version is *Species plantarum* published in 1753.[217] Linnaeus was aware of the artificial nature of his system, the value of which

207 Atran 1990: 158–161. On Jung, see Magnin-Gonze 2009: 102–104.
208 Sachs 1875 (Engl. transl. Garnsey 1890: 58).
209 Harvey-Gibson 1919: 27.
210 Harvey-Gibson 1919: 58.
211 Harvey-Gibson 1919: 20. This peremptory judgment is obviously not grounded in the text of *De plantis*.
212 Arber was the first woman botanist (third woman overall) elected as a Fellow of the Royal Society. Her famous historical works are her history of herbals (1912) and her *Natural philosophy of plant form* (1950).
213 Arber 1950: 31–32.
214 For example, Reeds' statements on Cesalpino's classification are mainly based on the works of Sachs and Arber (Reeds 1991: 19, 179).
215 Bremekamp 1952. Against Sachs' commentary of Cesalpino, see Morton 1981a: 157–158. Against Arber's unfounded criticism of Cesalpino, see Atran 1990: 151–153.
216 Assertions that appear to be made *a priori* in Cesalpino's theory are most often *a posteriori* (but presented as principled) theoretical (Aristotelian) justifications, grounded in empirical observations carried out over many years. Thus, Cesalpino's metaphysical positions are justifications rather than dogmas. Cesalpino tries to bring theory and practice closer together, rather than further apart. This will be discussed in the commentary.
217 Linnaeus 1753.

was primarily practical; but he harbored the ambition, never achieved, to reach a natural method of classification.[218] In this respect, Linnaeus was more influenced by Cesalpino than by any other botanist of the same period. Linnaeus admired Cesalpino's work and quoted him explicitly (e.g., in his *Genera plantarum*;[219] he attributes to him the famous quotation "*Confusis generibus omnia confundi necesse est*" ["All confusion is necessarily born of confused genera"[220]]).[221] According to Sachs, some sentences in Linnaeus' *Philosophia botanica*[222] and *Fundamenta botanica*[223] are copied from Cesalpino's text.[224] In addition, Greene shows that some tree genera identified by Cesalpino, but rejected by Linnaeus in 1753, were eventually re-established by 20th-century botanists, and concludes that "Cesalpino was, at least in so far, the true systematist and Linnaeus the superficialist."[225]

This is a very meaningful statement, considering the overwhelming influence of Linnaeus on botany and its history.[226] And yet, Greene's statement is debatable: although Linnaeus developed an artificial system of classification that made him famous, he was aware of the system's limitations. On this issue, Michel Adanson (1727–1806) refers to Cesalpino's natural classification, and analyzes it clearly, in more depth than Linnaeus had done.[227]

Below is a reproduction of Bremekamp's determination key, summarizing the taxonomic groups developed by Cesalpino (figures 5 and 6).[228] The table reconstructs the systematicity of Cesalpino's criteria clearly and synthetically:

218 Hoquet 2005; Müller-Wille 2005.

219 Linnaeus 1737: II.

220 Linnaeus (1737: II), quoting *De plantis*: Dedication letter IV.

221 Stafleu 1971: 66.

222 Linnaeus 1751.

223 Linnaeus 1736.

224 Sachs 1875 (Engl. transl. Garnsey 1890: 80–82).

225 Greene 1983: 2.811.

226 Greene considers that later botanists sometimes suggested classifications which were inadequate by comparison to Cesalpino's. He mentions the case of the walnut tree (of the genus *Juglans*), which Cesalpino correctly classified as a drupe fruit, even though authorities such as Augustin-Pyramus de Candolle (1778–1841) and his son Alphonse de Candolle (1806–1893) had – mistakenly – associated it with oaks (*Quercus*) and birches (*Betula*) (Candolle 1824–1873: 16.2, 134, 161). While Greene acknowledges the innovations of Cesalpino's classification, it must be stressed that Greene conceived his argument from a very positivist perspective of systematics. If one accepts that classifications are subject to constant reworking without being guided by an inherent tendency to progress, Greene's argument must be put into perspective.

227 Morton 1981a: 302–303.

228 Bremekamp 1952: 591–592.

Key to Cesalpino's groups

1*a*. Woody plants	
2*a*. with one-seeded fruits	
3*a*. with superior "blossom"	
4*a*. with coriaceous pericarp	group 1
4*b*. with osseous endocarp	group 2
3*b*. with inferior "blossom"	group 3
2*b*. with several-seeded fruits (eventually with pseudocarps containing several seedlike fruits)	
5*a*. without "blossom" or with superior "blossom"	group 4
5*b*. with inferior "blossom"	
6*a*. seeds inserted at the bottom of the fruit cavity	group 5
6*b*. seeds inserted on one or more longitudinal placentae	
7*a*. one placenta	group 6
7*b*. two placentae	group 7
7*c*. three placentae	group 8
7*d*. four placentae	group 9
7*e*. more than four placentae	
8*a*. fruit a cone	group 10
8*b*. fruit fleshy	group 11
1*b*. herbaceous and suffruticose plants	
9*a*. fruit unilocular	
10*a*. fruit one-seeded	
11*a*. fruit not crowned with a persistent calyx and not not enclosed within a perianth or involucre	group 12
11*b*. fruit crowned with a persistent calyx	group 13
11*c*. fruit enclosed within a perianth or involucre	
12*a*. rootlet turned away from the hilum or from the base of the fruit	group 14
12*b*. rootlet turned towards the hilum	group 15
add.: plants with a uninodal flowering shoot ("*Junci*")	group 15*a*
10*b*. fruit several-seeded	
13*a*. fruit a berry	group 16
13*b*. fruit a legume	group 17
α. plants with tendrils	
β. plants without tendrils	
13*c*. fruit with a central placentae	group 18
9*b*. fruit bilocular	
14*a*. fruit a double caryopsis	group 19
14*b*. fruit cells with a single free seed	group 20
14*c*. fruit cells with several seeds	
15*a*. dissepiment perpendicular to the plane of symmetry	group 21
15*b*. dissepiment in the plane of symmetry	group 22
9*c*. fruit trilocular	
16*a*. fruit a triple caryopsis	group 23
16*b*. fruit cells with a single free seed	group 24
α. plants with latex	
β. plants without latex	

Figure 5: Bremekamp 1952 table about Cesalpino's taxonomic groups 1/2.

16*c*. fruit cells with several seeds
17*a*. subterranean parts not bulbous . . . . group 25
17*b*. subterranean parts bulbous . . . . group 26
α. "blossom" inferior
β. "blossom" superior
add.: "*Liliaceae*" without bulbs . . . . group 26*a*
α. "blossom" inferior
β. "blossom" superior
9*d*. fruit quadrilocular or more than 4 fruits inside an involucre or perianth
18*a*. fruit quadrilocular (splitting into 4 nucules)
19*a*. rootlet pointing to the top of the nucule . . . group 27
19*b*. rootlet pointing to the base of the nucule. . . group 28
18*b*. more than 4 fruits inside an involucre or perianth
20*a*. each fruit originally provided with "blossom" group 29
α. "blossom" on all ruits ligulate
β. "blossom" marcescent
γ. "blossom" ligulate on the marginal fruits only
20*b*. fruits with a common "blossom" (i.e. formed inside a perianth)
21*a*. fruits with a single seed . . . . group 30
21*b*. fruits with several seeds . . . . group 31
9*e*. without fruit and seed . . . . group 32
α. plants provided with root and shoot
β. plants without root and shoot.

Figure 6: Bremekamp 1952 table about Cesalpino's taxonomic groups 2/2.

Cesalpino's distinction between woody and herbaceous plants is borrowed from Theophrastus. The innovative character of his method lies in the inclusion of secondary grouping characteristics, based on very precise observations of flowers, fruits, and seeds. For example, is the flower hypogynous or epigynous? Does the fruit remain attached to its calyx? Is it unilocular, bilocular, etc.? Many of these criteria were first studied by Cesalpino and are still relevant in current systematic botany.

Below is Bremekamp's diagram linking these taxonomic groups to the families they contain (figure 7):[229]Cesalpino successfully identified groups corresponding to modern families such as Fabaceae, Liliaceae, Lamiaceae, Apiaceae, Poaceae, Chenopodiaceae, Brassicaceae, Rubiaceae, Scrophulariaceae, Euphorbiaceae, Renonculaceae, Asteriaceae, Valerianaceae, Primulaceae, Caryophyllaceae, Verbenaeaceae, Cucurbitaceae, Boraginaceae, and Rosaceae.[230] Considering how natural these groupings appear, the comments of Sachs, Harvey-Gibson, and Arber on the artificiality and apriority of Cesalpino's criteria should be tempered. Bremekamp writes:

229 Bremekamp 1952: 593.
230 Magnin-Gonze 2009: 80.

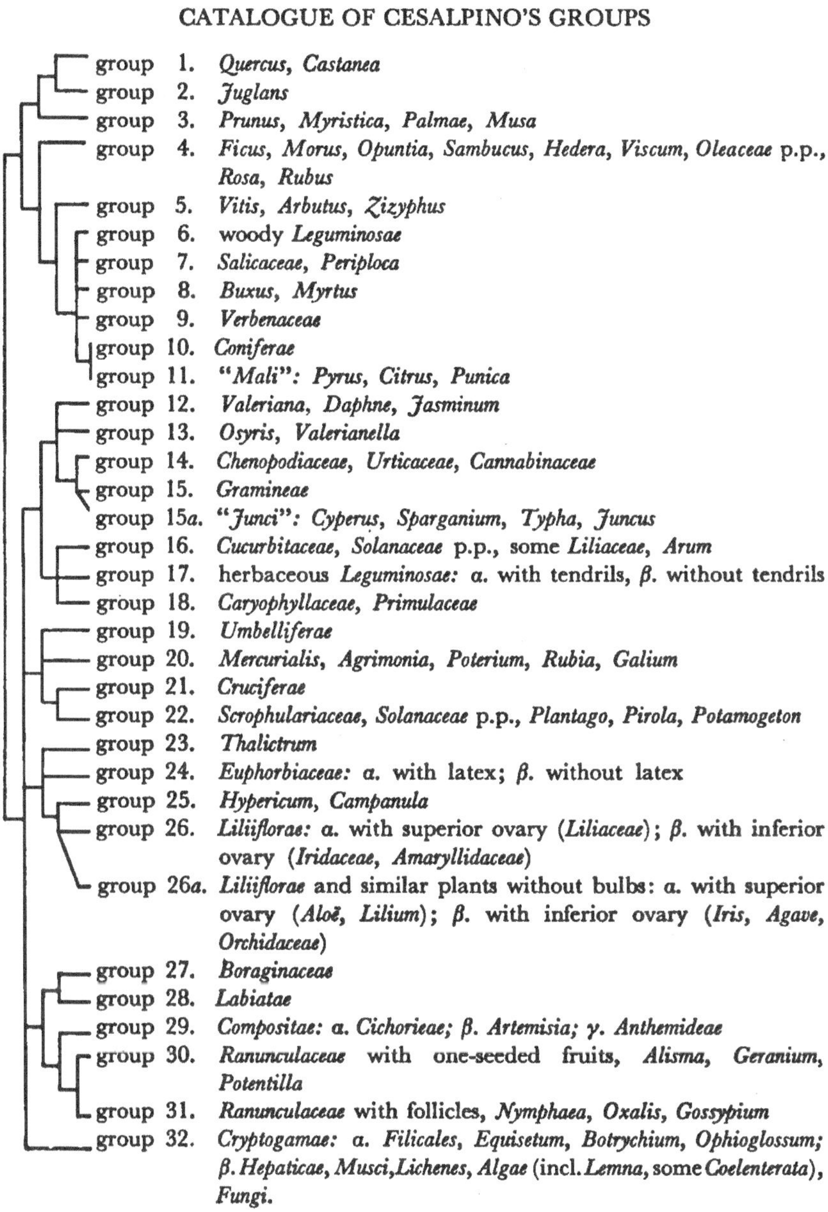

Figure 7: Bremekamp 1952 diagram linking Cesalpino's taxonomic grouping with plant families.

> In judging the trials and errors of these pioneers in the domain of taxonomy, we should not forget that the need for such a "logical" justification was felt very strongly in their time, and that without this addition, which to us is hardly more than a literary embellishment of somewhat doubtful taste, their systems would not have been acceptable to their contemporaries. The artificial nature of the arguments by means of which Cesalpino tried to make the choice of his characters acceptable, makes it probable that these characters were not, as it is customary to assume, chosen on account of a preconceived notion, but that they must have been brought to light by the analysis of natural groups, i.e., of groups of plants that are similar in habit.[231]

Cesalpino's innovations are all the more significant that Aristotelian doctrine prevailed in the 16th century as an authority almost as strong as the Bible. Cesalpino's classification is certainly open to criticism, but the choice of his morphological criteria was likely the result of a great deal of observation rather than pure deduction. Moreover, molecular systematics and cladistics of the late 20th and 21st centuries have called into question the evident naturalness of certain morphological groupings used by botanists like Sachs and Arber to assess their predecessors.

In contrast to Bremekamp's analysis, Atran suggests that the aforementioned families identified by Cesalpino "were, to a significant extent, already known to local folk and the herbalists of the time."[232] Atran posits that Cesalpino relied on trial an error to group plants, considering successive sets of fructification characteristics, until he reached a satisfactory reconstruction in agreement with intuitive groupings, and "occasionally" forming new ones.

In any case, and despite the modernity of its innovations, the classification suggested in *De plantis* remains dependent on Theophrastus' historical division into four main genera, which have little relevance in present day botany.[233] The characteristics of fruits and seeds are accurate and useful to group plants, but Cesalpino does not specify which should be given priority in classification and diagnosis.[234] Cesalpino was also largely guided by finalist ideas; the gradation of the chain of beings can be detected in his organization of plants according to a finalist order of perfection.[235] In books II to XVI, *De plantis* adopted a classificatory order of presentation from the higher genera (glandiferous trees) to the lower plants, in parallel with the morphological grouping of plant genera. Non-vascular plants and more generally cryptogams are thus considered the most imperfect (see chapter 14 for the justification of this categorization).

---

231 Bremekamp 1952: 584.

232 Atran 1990: 156.

233 From the point of view of the current phylogenetic systematics of species, there is nothing to prevent, for example, a tree and an herbaceous plant from belonging to the same genus, following differentiation from a common ancestor. This situation explains why the two species share important characteristics.

234 Atran 1990: 157.

235 On the chain of beings, see Lovejoy, 1936, especially chapter IV on the influence of the idea in Renaissance science.

Finally, Cesalpino's classification nomenclature is not conceptualized. The very idea and definition of the modern genus as a group of similar species remain implicit in *De plantis* (like the modern family taxon), even though his conception of the genus and his use of particular genera were very close to the current notions.[236] To define the concept of species, Cesalpino relied on reproductive criteria: a plant belongs to a species when it has been produced by a seed of this species.[237] However, identifying and classifying plant species (beyond the general concept of species) requires the identification of specific morphological characteristics.

Status of the Illustration in *De plantis*

De plantis is not illustrated. This is surprising, since most contemporary botanical treatises included (more or less accurate) images of plants.[238] There are several possible reasons for this absence of illustrations. Finding suitable artists, and the cost of having the engravings made and printed for the sixteen books may have been a deterrent, although this hypothesis is purely contingent, since Cesalpino received financial support from patrons, and since illustrations were commonly reused.[239] And yet not a single drawing or diagram, however inexpensive, appears in the treatise. Another reason may be Cesalpino's endorsement of a medieval epistemic conception of images. According to contemporary historiography, illustrations in the Middle Ages would often have been perceived as inferior to text and ideas and associated with illiteracy, imprecision of meaning, lack of rigor, and the dissemination of knowledge outside the sphere of the initiated.[240] Indeed, authorities of medical botany, such as Dioscorides, Galen, and Pliny the Elder, may have refused to use illustrations,[241] a stance that Fuchs was among the first to criticize in the 16th century.[242] Rather, the modernity of Cesalpino's descriptions and his dissociation from pharmacotherapeutic tradition suggest that the absence of illustrations in *De plantis* is a methodological choice highlighting the potency of the descriptions. Jean Ruel, one of Cesalpino's probable sources of inspiration, had adopted the same strategy.

236 Dughi 1957: 141.

237 See §133 and our commentary.

238 Arber 1912; Givens et al. 2006; Swan 2006, 2011; Ogilvie 2006.

239 For example, Dodoens' *Cruydeboek* (1554) reused Fuchs' engravings and a latter augmented edition of his herbal called *Stirpium historiae pemptades sex sive libri triginta* (1583) included illustrations from the works of Charles de l'Écluse and Mathias de L'Obel, because they shared an agreement with their common editor, Christophe Plantain (1520–1589) (Arber 1912: 72–74; Swan 2011: 189).

240 Magnin-Gonze 2009: 32–33.

241 Beretta 2001.

242 Swan 2011: 190.

In the 16th century, botanical illustrations became marked by naturalism and a concern for detail, unlike most medieval representations.[243] Otto Brunfels' *Herbarium vivae eicones*[244] is often cited as the driving force behind this trend.[245] However, despite the accuracy of their illustrations, these herbaria do not always contain precise textual descriptions of morphology and anatomy, and rarely adopt a convincing method of classification.[246] And yet, one or two lines of description of a species may be just as accurate as a full-page image, especially at a time when painters and engravers often took liberties. The absence of illustrations in *De plantis*, compared to treatises of the same period, may be explained by its descriptive rigor and its innovative natural classification by affinity of genera. Indeed, as plants belonging to the same genus have many common characteristics, only a few descriptive elements are necessary to identify species once their genus is given. Thus, referring to Theophrastus, Cesalpino explains in the preface of *De plantis*:

> Qui autem secundum naturam societatem assignatur, omnium facillimus reperitur, tutissimus, vtilissimusque tum ad memoriam, tum ad facultates contemplandas: [...] In persequenda autem secundum hunc ordinem plantarum historia fit, vt breuior descriptio satisfaciat, non enim cogimur in singulis ea repetere, quae communiter generibus conueniunt: atque adeo certa ex hac breui descriptione paratur notitia, vt pictura certiorem efficere non possit: Non enim omnis differentias pictura exprimit, vt oratio.
>
> He who assigns them a classification according to nature, finds himself at the greatest ease, security, and advantage of all for memorising as well as observing [their] properties. [...] The inquiry pursued according to this way of ordering plants has the effect that a shorter description is sufficient, for we are not forced to repeat for individual [plants] what is common to the whole genera; and thus is gained from this short description such a solid knowledge that a picture could not produce a more certain one: indeed a picture does not show all the differences, as words can.[247]

He also voices his reservations about the reliability of illustrations:

> [...] purissimam scilicet stirpium historiam continere, nullis figmentis adulteratam, qualem saepe in impressis picturis inspicimus.
>
> [my work will] contain a very clear enquiry on plants, unadulterated by inventions, as is often observed with printed pictures.[248]

Cesalpino believed that images could be misleading, and less accurate than good descriptions. If a genus is clearly described, a new species of the same genus, even if unknown

---

243 Swan 2002.

244 Brunfels 1530.

245 On Brunfels, see Greene 1909: 169–191; Arber 1912: 47–50, 168–174; Ogilvie 2006: 193–196; Magnin-Gonze 2009: 62–66.

246 Arber 1912; Swan 2011: 187.

247 Cesalpino 1583: Dedication letter V. See also Heideklang 2023.

248 Cesalpino 1583: Dedication letter VII.

to the reader, could easily be identified, or even represented, by a very short description. Cesalpino was aware of the descriptive precision resulting from his system, and it was this system, as well as his predictive methodology, that led him to reject illustrations as non-essential.[249] This approach to knowledge generated the binomial Linnaean nomenclature: once the genus has been determined, a single term – the epithet – added to the genus name is sufficient to refer to a species. The prevalence of short and precise descriptions on images certainly fostered the evolution of (binomial) nomenclature.

To remain faithful to Cesalpino's choice, our edition of the translation is not accompanied by images. All illustrations added for pedagogical purposes are included in the commentary section.

The Dried Herbarium of Cesalpino (*Hortus siccus*)

Cesalpino is also known as the author of one of the first dried herbarium, created between 1553 and 1563.[250] This is not surprising since his master, Luca Ghini, is credited as the inventor of the technique. Cesalpino's herbarium is currently one of the six oldest in existence, along with Aldrovandi's and Michele Merini's (ca. 1540–1545). The origin of the three others is less certain. One, dated to 1532, is attributed to another student of Ghini, Gherardo Cibo (1512–1600), or more likely Francesco Petrollini (*fl.* mid-16th century). Petrollini may also be the author of the anonymous *En tibi* herbarium (ca. 1558) according to recent studies. The sixth is the *Estense herbarium* (ca. 1570–1600) which remains anonymous.[251]

19th-century botanist Théodore Caruel (1830–1898)[252] inventoried, identified, and described the species in Cesalpino's herbarium.[253] Of modest size (768 specimens on 260 sheets corresponding to 760 species)[254] compared to Aldrovandi's herbarium of the same period (about 5000 plants on more than 4000 sheets), Cesalpino's herbarium has often been considered superior from a scientific point of view, because it is the first to

249 On the dedication letter of *De plantis*, see also Baldassarri 2022. On the debate regarding the epistemic value of pictures in contrast to words, see Kusukawa 2012; Olariu 2018. Heideklang (2022) notes, however, that Cesalpino's stance in *De plantis* is not clear, since he suggests that he might have printed a luxury edition with woodcuts, had he received adequate funding. However, the idea that images are non-essential and sometimes inaccurate does not imply that they must be wholly rejected. On that epistemic stance, Cesalpino appears subtler than the Ancients simply refusing illustrations.

250 On the history of herbaria in general, see Arber 1912; and Thiers 2020: 18–19, on Cesalpino's herbarium.

251 Stefanaki et al. 2019.

252 Caruel 1858.

253 Saint-Lager 1885.

254 Moggi 2009.

compile plants according to a classification.[255] Two copies were made: one was given to Cosimo I, Grand Duke of Tuscany, and is now lost. The other, given to Alfonso Tornabuoni (1504–1577), Bishop of Florence, around 1563, has survived along with a very interesting letter of dedication.[256] This herbarium is currently kept in the Botanical Museum of Florence where it was restored and rearranged in 2003, for the 400th anniversary of Cesalpino's death.[257]

This herbarium provides invaluable insights into the botanical conceptions of its author, for two reasons. Firstly, because it is organized according to the principles of classification developed in *De plantis* twenty years before its creation,[258] thus demonstrating that Cesalpino's classification had long been governed by a great sense of observation and long-term empirical rigor. This invalidates the claim that Cesalpino's system was based on strictly *a priori* criteria and vision. Secondly, the specimens in the herbarium allow modern botanists to identify a large number of plants mentioned in *De plantis* under sometimes unrecognizable names.[259] The herbarium also lists the Latin, Greek, and Italian names of each specimen. By studying Cesalpino's herbarium, Théodore Caruel was able to identify most of the species for which he provided the equivalent names in the binomial nomenclature of the time in an index published in 1858.[260] This greatly helped us to identify the species of Book I of *De plantis* from the Latin names of 1858, and thus to establish our own index with the current binomial terms.

The fact that Cesalpino was a herbalist and collected specimens himself[261] also sheds light on the problem of the lack of illustration. The real plant, even as a dried specimen, holds an epistemologically higher status than its representation (even from nature). At a time when botanical gardens were first developed, the materiality of the real plant took precedence over representation in the teaching of botany. Although illustration techniques evolved simultaneously, gardens made it possible to learn and study while observing real plants; herbariums even allowed this at any time of the year. For the first time, herbariums enabled botanists to compare a wide variety of specimens in the same place, but also at the same time. Real specimens and even whole herbariums could be exchanged with, or passed down to, other botanists. Illustrated botanical works attesting to the improvement of the accuracy of illustrations remain exception-

---

255 Saint-Lager 1885: 45; Moggi 2009.

256 The letter is published in Caruel 1858: 1–6. For the history of the herbarium and its transmission, see: Moggi 2009. On the letter, see Morton 1987.

257 Nepi 2007; Nepi and Gusmeroli 2008.

258 Bellorini 2016: 71.

259 Since *De plantis* describes about 1500 species and the herbarium only 760, several species (often common or cultivated species) mentioned in Book I do not appear in the herbarium. For these cases, see our index of plant names.

260 Caruel 1858: 7–24.

261 See the dedication letter for references to his expeditions.

ally rare, whereas the majority of drawings in books on plants remained of mediocre quality until the 18th century.[262] Real specimens thus became the simplest, most reliable, and least expensive point of reference for early modern plant studies, although not until the second half of the 16th century. It is on these issues that Luca Ghini's influence on Cesalpino is the most probable.

The Physiology of Cesalpino

In addition to his classification innovations, Cesalpino is also known for his theoretical reflections on plant physiology. Although his ideas on this subject may not have been directly decisive for the development of the natural sciences, they are important in the history of science, as they pave the way for personal observations and hypotheses. Cesalpino drew on his knowledge of animal anatomy and physiology, as well as on physics, to propose new ideas on the functioning of plants, and to distance himself from some of the theses of the Ancients.[263] He thus compared the animal circulatory system to the vascular system of plants,[264] and elaborated physical hypotheses to explain the movement of sap in the context of plant nutrition.

As for growth and development, he proposed what we call in our commentary the "theory of interlocking layers" to link his physiological speculations on plant nutrition, circulation, and growth with his knowledge of plant anatomy. According to this theory, each layer of the anatomy of the fruit, seed, or flower originates in one of the three main layers of the stem (or trunk), namely the pith, wood, and bark. This correspondence between the tissues of the different parts suggests that they share a common nature from which similar qualities may be deduced. This theory, largely deductive, has now been disproved, but it had a profound influence on Linnaeus' ideas on flower generation in his *Classes plantarum* and *Systema natura*.[265]

> Theophrastus and Cesalpino maintained that basically a plant consisted of a *medulla* (which was the seat of its essence) and a *cortex*. This cortex could consist of xylem, phloem (*liber*), cortex in sensu stricto and the epidermis. Linnaeus held that in the flower the calyx represented the cortex, the corolla the phloem, the stamen the xylem (and therefore part of the cortex!) and the pistil the medulla: the ovules were the true carriers of the essence (*Syst. Nat.* ed. 10. 2: 826).[266]

262 Arber 1912: 199–203.
263 Morton 1981a: 134; Hiernaux and Tresnie 2023a.
264 On this idea, and more generally on Cesalpino's contribution to the discovery of blood circulation, see the commentary.
265 Sachs 1875 (Engl. transl. Garnsey 1890: 49–51).
266 Stafleu 1971: 137.

Regarding plant reproduction, Cesalpino's ideas remained traditional. He did not really envisage sexual reproduction in plants. We comment more specifically on these passages in our systematic analysis of the text. While Cesalpino rigorously observed certain details and indeed valued induction, he did not engage in practical experimentation or establish a scientific protocol; rather, he resorted to observation and thought experiments. Thus, although he may have tested some of his ideas empirically, many of his arguments seem purely deductive and aprioristic. Cesalpino's observations were also limited by the technology available at his time, notably by the absence of a microscope. This contrast between medieval thought, its ancient sources, reasoning by analogy and deduction, and the slow advent of early modern science mirror Cesalpino's situation. A step towards early modern science is made through the use of direct observation, through the importance of the empirical and inductive method, reflected in the mechanical explanations and analogies used to describe plant nutrition, and through the criticism of authorities and of popular traditions.

Part II: **Latin Text and Facing English Translation**

https://doi.org/10.1515/9783111001104-003

Notes on Editing

We offer here a transcription of the first book of the original printed edition, published by Georgius Marescottus (Giorgio Marescotti, *fl.* 1563–1601) in Florence in 1583. Page numbers of this edition are indicated in the left margin of the text in square brackets. While we have preserved the division into chapters, titles and paragraphs – marked in the text by a section sign <number> – are not present in the original text but adopted here to facilitate the reading of the text and its facing translation. This structure follows the logical articulations of the text, only diverging from its original punctuation in the few cases when preserving it would have been misleading. Line numbers added in the right margin do not correspond to the Latin lines of the orginal edition.

Our main intervention in the original text was to apply the fourteen substitutions, additions, or deletions, either of a word or of a punctuation mark, listed by Cesalpino himself in the erratum included at the end of the original edition. These corrections are signaled in footnotes by the line number and "C. corrigit". The resulting text retains thirteen typos or curiosities, which we have amended, and indicated in footnotes as "corrigimus" preceded by their line number. Secondly, we have chosen to resolve the following abbreviations and ligatures: *-q* becomes *-que*; *&* becomes *et*; *æ*, *œ* and *ę* become *ae* or *oe* as appropriate; the raised full stops and dashes become *m* or *n* as appropriate. In addition, *ſ* is normalized to *s* and the grave accents sporadically used on prepositions, adverbs and ablatives are removed. Excepting the corrections from Cesalpino's erratum, these interventions do not modify the meaning of the text; they only serve the comfort of the modern reader.

We have otherwise faithfully reproduced the printed edition of 1583. We do not systematically underline deviations from classical grammar; Cesalpino, like many of his philosophical and scientific contemporaries, retained several habits from scholastic Latin, in spite of the attempts of Italian philologists to reform the language in the 15th century. Moreover, the original edition does not use capital letters consistently. Words may be capitalized after full stops, colons, and semicolons. Names of genera, subgenera, and species are usually, but not always, capitalized. Names of plant parts are not usually capitalized, with some exceptions. The title words of Cesalpino's *Quaestiones peripateticae* are sometimes both capitalized, sometimes not. Finally, in two cases, Cesalpino writes the same word in two different ways: *Arundo* and *Harduno* for reed, *Cyclaminus* and *Ciclaminus* for cyclamen. Since these irregularities do not hinder reading, we have left them as they are.

Notes on the Translation

Translation choices were made with the intention to render Cesalpino's text as clearly as possible without altering its nuances. Firstly, for the parts of the plants and phenomena described by Cesalpino, we chose the modern terms closest to Cesalpino's descriptions, while refraining from adding anachronistic technical precision to the text. Our choices sometimes depart from more usual translations of botanical terms. To avoid ambiguity, the first occurrence of each technical term and each species name is accompanied by the original Latin term, and the same translation is then retained throughout the text. The only exception is the term *germinatio* – see our comments to chapters 3, 4, and 6. Some non-technical expressions are also quoted in Latin between brackets for the sake of accuracy.

As Cesalpino's Latin is sometimes elliptical, we have occasionally supplemented in the translation words between square brackets. Similarly, the titles of chapters and paragraphs are ours, and are therefore also displayed between square brackets. Our translation does not always follow the original punctuation: Cesalpino's sentences are very long, incident propositions are not always clearly marked, and sometimes do not correspond to the logical progression of the argument. Our own division into sentences and the addition of brackets are intended to make Cesalpino's argument clearer. They are therefore part of our translation and interpretation of the text.

Cesalpino almost always capitalizes the names of species, often also those of subgenera. From chapter 12 onwards, where the four main genera (tree, shrub, undershrub, and herbs) are properly defined, he also capitalizes their names. We did not follow this use in the English translation, as the Latin text is not consistent on this point. For the rest, we normalize the above-mentioned Latin irregularities. We indicate our other occasional interventions in footnotes. The Latin text does not contain notes. All footnotes are ours; they provide explanations about the translation, as well as references or short explanations on species identification, or on Cesalpino's ideas developed further in our commentary.

Table of Chapters

Chapters	English title	Pages in the Latin edition	Paragraphs in the English translation
1	The properties and parts of plants	1–3	1–14
2	Nutrition	3–5	15–26
3	Development and growth: embryo, bud, shoot, and bark	5–8	27–37
4	Growth and development: heart, stem, soboles, and leaves	8–10	38–47
5	Vegetative reproduction	10–11	48–54
6	Seeds	11–14	55–63
7	Flowers	14–17	64–76
8	Fruits and seed coats	17–18	77–85
9	The pericarp	18–20	86–92
10	The parts dedicated to the protection of the fruit and fructification	20–23	93–113
11	The secondary parts	23–24	114–117
12	The four main genera and their divisions	24–26	118–134
13	Criteria for defining subgenera and species	26–28	135–144
14	Subdivisions based on the reproductive operation	28–30	145–157

[1] ANDREAE CAESALPINI ARETINI
DE PLANTIS LIBER PRIMVS

CAP. PRIMVM.

<1> Cvm natura plantarum illud solum genus animae sortita sit, quo alantur, crescant, et gignant sibi similia; careant autem vi sentiendi, mouendique in quibus animalium natura consistit: iure optimo plantae longe minori instrumentorum apparatu indiguerunt quam animalia. <2> Multae enim sunt partes in animalibus ad sensum facientes, quae multiplici forma et numero sunt constitutae: plures adhuc in eisdem reperiuntur ad motus praestandos: nam huius gratia vniuersa fere ossium substantia articulis distincta est, et caro in musculos est digesta neruis in omnes partes discurrentibus; si praeterea viscera contemplemur, quae altricis animae sunt instrumenta, ob similem animae facultatem, modicam quidem similitudinem cum plantarum partibus intuebimur: sed in plerisque maximam dissimilitudinem. <3> Natura enim venarum, quae alimentum ex ventre hauriunt, vt illud in vniuersum corpus distribuant; aliqua ex parte respondere videtur cum plantarum radicibus; nam similiter hae ex terra tanquam ex ventre cui implantantur trahunt alimentum. Cum autem animalia exquisitiori ciborum genere indigerent, et multa eorumdem praeparatione coctioneque eorum radicibus ventres sunt appositi, aliique multi ductus ad alimenti excrementa separanda. Quae omnia plantis negata sunt. <4> Quo fit, vt corpora plantarum valde simplici substantia constare videantur, et maxime accedere ad naturam similarium, discedere autem ab operosa organicarum compositione. <5> Quoniam autem altricis animae

Chapter 1 [The properties and parts of plants]

1. [Faculties of the soul]

While the nature of plants has been endowed with the same kinds of soul [as animals] (through which they feed, grow and reproduce identically), [plants] still lack the capacity to feel and to move, of which the nature of animals consists. Plants are accordingly deficient in comparison with animals: they are much less endowed with faculties.

2. [Faculties of animals]

Indeed, in animals, feeling is made possible by many parts of the body, which vary in shape (*forma*) and number. Several of these also contribute to movement. To that end, the substance[267] of the bones is almost always separated by joints, and the flesh distributed in the muscles by nerves[268] which run through all parts [of the body]. Furthermore, if we consider the viscera, which are the feeding tools of the soul (that is to say they serve the nutritive faculty of the soul), we can only find a slight resemblance to the parts of plants, while in many places, we can see a great dissimilarity.

3. [Nutrition]

However, the nature of veins, which bring food from the abdomen (*venter*) and distribute it throughout the body, seems to partly correspond to the roots of plants. Indeed, the two embed themselves in the earth and in the abdomen in a similar way, and draw food from them. In contrast, animal species need a refined kind of nutrients: their entrails (*ventres*) (from their "roots"[269]) serve to make many of these foods available to them and to process them, while others [entrails] act as conduits to separate the excrement from the food. None of this applies to plants.

4. [The simplicity of plants]

As a result, the bodies of plants seem to remain firmly in their simple substance and to most closely resemble the nature of similar [bodies], but to distance themselves from the industrious arrangement of the organs.[270]

267 *Substantia*, for Cesalpino, as is the case throughout the Aristotelian tradition, sometimes indicates the base material, sometimes the essence, and sometimes both at the same time. See our introduction on this point.

268 *Nervus* can refer to any form of nerve, cord, tendon or joint.

269 Here and henceforward, we have added quote marks to indicate that Cesalpino uses the term in an analogical sense: here he is talking about veins, which are like roots without really being so.

270 It is unclear what these similar bodies are, but the argument might be that plants are similar to animals by the simplicity of their bodies, but dissimilar as animals have more complex organic systems.

opus est gignere quale ipsum, siue id ex alimento fiat ad conseruationem singulorum, siue ex semine ad specierum aeternitatem: duae ad summum partes perfectioribus datae sunt, maxime necessariae: Vna, per quam sumant alimentum, quae radix appellatur. Altera qua fructum tanquam foetum ad speciem propagandam ferant, qui caulis vocatur in humili materia, caudex autem in genere arboreo. <6> illa quidem
[2] superior, quia principalior licet intra terram condita sit; viuunt enim multae plan|tae sola radice, postquam caulis exaruerit perfecto semine, vt Cyclaminus, Aristolochia, et pleraque inter Acanacea, et Ferulacea; haec vero inferior quamuis supra terram attollatur; excrementa enim, si quae sunt, per hanc partem excernuntur: quomodo in animalibus pars superior et inferior accipitur. <7> Verum si nutritionis modum inspexerimus, alio modo partem superiorem inferioremque statuemus. Cum enim tam in animalibus quam in plantis nutrimentum sursum feratur; leue enim est, quod nutrit, vtpote quod a caliditate sursum ducitur: radices in inferiori parte implantari necesse fuit, caulem autem recta sursum attolli: nam et in animalibus venarum radicatio in inferiori ventre est, caudex autem sursum petit ad cor et caput. <8> Vtrum autem aliqua pars in plantis statuenda sit, in qua sit animae principatus, vt cor in animalibus, considerandum est; Esse igitur in illis oportere aliquam huiusmodi partem ex eo coniectari licet. Cum enim anima sit actus corporis organici, neque tota in toto esse potest, neque tota in singulis partibus, sed tota in aliqua parte principali, vnde caeteris ab ea pendentibus vita communicetur, vt vniuersaliter a nobis ostensum est in Quaestioni-

5. [The nutrition and generation of plants]
But since an operation of the nutritive soul is to reproduce identically (which it does either through food for the preservation of the individual, or through the seed (*semen*) for the continuation of the species), plants must necessarily be endowed with two main parts to be complete. One of these, through which they procure food, is called the root; the other, through which they bear both fruit and progeny (*fetus*) in order to propagate the species, is called a stem in small plants, and trunk in the genus of trees (*arbor*).

6. [The superiority of the root over the stem]
The first part, although more important and therefore superior, is hidden in the earth. Indeed, many plants survive | only with their root, once their stem has withered, after [2]
their seed has been produced, for example the spring sowbread (*Cyclaminus*), the smearwort (*Aristolochia*), and many other species including those in the Acanaceous (*Acanaceum*) and the Ferulaceous (*Ferulaceum*) genera. The second, although it emerges above the ground, is in fact inferior: it is through this part that the excrements, if there are any, are evacuated. This is how we consider the superiority or inferiority of parts of an animal.

7. [The inferiority of the lower part of the body]
Actually, if we look at the mode of nourishment, we will view the superiority or inferiority of each part in a different way. In fact, as much in animals as in plants, the food is transported from the bottom up (that which nourishes is in fact light), because that which heat carries is from bottom to top: it was therefore necessary for the roots to be embedded in the lower part and for the stem to rise straight upwards. In animals too, the veins (*vena*) are "rooted" in the lower part of the abdomen, while the "trunk" goes straight upwards, from bottom to top, towards the heart and the head.

8. [The location of plants' soul]
We must ask ourselves if it is necessary that a place exists in the plant where the principle of the soul is situated, as the heart in an animal; it is in any case legitimate to conjecture that it would be fitting that such a place should exist. Although the soul is the actuality of a body which has organs[271], it can neither be entire everywhere, nor entire in several unique parts, but must be entire in one chief part, from where life is distributed to the other dependent parts, as we have shown in a universal context in *Quaestiones peripateticae*.[272]

271 Aristotle, *De anima*, 412a27 (ed. Ross 1961: 26).
272 Cesalpino 1571: V, 7 (see also I, 7 and V, 3). On Aristotle's position and arguments on the subject of the soul and the divisibility of the body of plants (and of certain worms or insects), see *De anima*, 409a5–10, 411b9–30, 413b16 (ed. Ross 1961: 18–29); *De juventute et senectute*, 2, 468a28–31 (ed. Ross 1955: 468); *Metaphysica*, Z, 16, 1040b5–15 (ed. Jaeger 1957: 161–162); *De Incessu animalium*, 7, 707a27 sq. (ed. Falcon and Stavrianes 2021: 59).

bus Peripateticis: <9> ex propriis autem, quae circa plantas contingunt, hoc modo ostendi possit; si enim radicis opus est alimentum ex terra trahere: caulis autem semen ferre: neque commutantur vnquam, ut radix semen ferat, et germen intra terram condatur; aut duae erunt animae specie distinctae et loco separatae, vt altera in radice sit, altera in germine: aut vna erit vtrique propriam facultatem impartiens: <10> Non esse quidem duas animas specie diversas, et loco separatas in vna planta illud argumento est: quod saepe videmus radicem abscissam germen emittere, et ramum abscissum radicem intra terram egerere: quasi indiuisibilis specie anima sit in vtrisque partibus. Hoc autem videretur ostendere totam animam in vtrisque partibus esse, et totam in tota planta, nisi illud obstaret, quod similiter in multis videmus, diuisas esse omnino facultates in vtrisque partibus, vt germen, quomodocunque fodiatur, radices nunquam emittat, et si abscindatur, pereat: vt in Pinu et Abiete, radices quoque auulsae in eisdem pereunt: quasi principium quoddam sit vitale, a quo vtrisque vita impartiatur, quamdiu adnatae illi fuerint. <11> Verum enimuero difficile fuerit in plantis huiusmodi partem inuenire, in qua sit animae principatus. Si enim illas plantas inspexerimus, quae multum temporis viuunt sola radice, postquam caulem cum semine abiecerint, videbitur principium in radice esse, si vero rursus illas intuebimur, quae ex ramo, aut surculo propagantur, vt Olea, Vitis, et Punica, principium in germine esse fatebimur; Erumpit enim ex illis radix, si serantur. <12> Si praeterea partium differentias contemplemur, tum in radice, tum in germine, in radice quidem omni duas partes
[3] inspiciemus, corticem scilicet, et corpus, quod intra corti|cem continetur; in aliis quidem durum lignosumque; in aliis autem molle carnosumque: in caule autem tria partium genera ipsum constituentia, corticem scilicet externum, medullam internam, et medium corpus inter corticem et medullam, quod in arboribus lignum vocatur.

9. [The unity or multiplicity of the soul in plants]
With regard to the particular case of plants, we can illustrate it in the following way. If the operation of the root is indeed to extract food from the soil, and that of the stem to carry the seed, the two can never exchange roles, with the root transmitting the seed and the shoot (*germen*) embedding itself in the earth. Therefore, either there are two souls, distinct in species and separate in location, such that one is in the root and the other in the stem; or there is one single soul distributing to each of the two parts its own faculty.

10. [Neither two separate souls nor one single soul present everywhere]
One argument challenges the idea of two souls distinct in species and separate in location within one plant, since we often see a shoot growing from a cut root, and a cut twig (*ramus*) taking root in the soil: as if a soul, indivisible in its kind, were revealing itself in both parts. This would seem to indicate that the soul in its entirety is present in each of the parts, and in its entirety in the whole of the plant, if it were not for the fact that in many plants, we can see that the faculties are completely distributed between their parts. For example, the shoot, wherever it is pierced in any way, never puts down a root, and in the case of the pine (*Pinus*) and the fir (*Abies*), it dies if it is severed. In these same species, any roots that are cut off also perish; they form a kind of principle of life, from which life is distributed to each of the parts, as long as they remain attached.

11. [The difficulty of locating the soul of plants]
In reality, however, it would be difficult to locate in plants one such part which contains the principle of the soul. Indeed, if we examine plants which survive a long time with only their root, having lost their stem with the seed, the principle seems to be found in the root; if on the other hand we were to look at those that propagate from the twig or the scion (*surculus*),[273] for example the olive tree (*Olea*), grapevine (*Vitis*) or pomegranate (*Punica*), the principle would be found in the shoot: the root emerges from it, if they are planted.

12. [The anatomy of the root and the stem]
Let us look at the differences within the parts, whether that be in the root or in the shoot. In all roots, we can observe two parts, namely the bark (*cortex*) and the body (*corpus*) | that it contains: in some cases, it is hard and woody, in others tender and [3]
fleshy. The stem, on the other hand, is made up of three parts, namely the bark, the internal pith (*medulla*)[274] and the body between the bark and the pith, which in the case of trees is called "wood" (*lignum*).

273 A scion is a one-year-old shoot of a plant.

274 Cesalpino uses the same word – *medulla* – to refer to the pith of plants and the brain marrow of animals.

<13> Si igitur in omnibus natura in intimis recondere solet vitalia principia vt viscera in animalibus: rationi quoque consonum fuerit in plantis non statim intra corticem, sed intimius principium condidisse, scilicet in medulla interna, quae tantum in caule est non in radice. Hanc autem fuisse antiquorum sententiam ex nomine coniectari possumus: hanc enim partem in plantis cor appellabant, alij Cerebrum, alij matricem, quasi hinc faetificationis principium ducatur; Quoniam autem in medio esse maxime oportuit id quod exteris partibus vitam distribuit. duae autem sunt partes plantarum maxime conspicuae radix scilicet, et id totum, quod sursum attollitur, merito in intermedio, qua scilicet radix germini coniungitur, locus videatur cordi plantarum oportunissimus. <14> Apparet autem in hoc loco substantia quaedam, tum a germine, tum a radice distincta, mollior enim et carnosior est vtrisque, unde cerebrum appellari solet, in multis cibo accomodata antequam senescat; nam per aetatem dura et lignosa redditur, vt caeterae quoque partes, pulcherrime autem huic parti conuenire videtur cerebri nomen, quemadmodum enim in animalibus cerebri medulla in capite est, vnde spinalis medulla exoritur in totam spinae longitudinem diducta, sic in plantis cerebrum in radice tamquam in capite sedens per totum caulem quasi per spinam dorsi medullam deducit ad vitalem humorem ramis et extremis surculis distribuendum.

CAP. II.

<15> Qva autem ratione fiat alimenti attractio, et nutritio in plantis, consideremus. Nam in animalibus videmus alimentum per venas duci ad cor tanquam ad officinam caloris insiti, et adepta inibi vltima perfectione per arterias in vniuersum corpus distribui agente spiritu, qui ex eodem alimento in corde gignitur. In plantis autem neque

13. [Locating the principle of life in pith and root collar]
If, in all things, nature tends to be embedded in the internal parts, the principle of life, like the intestines in animals, in harmony with reason, will not simply be in the bark, but rather more deeply embedded in the inner pith, which is found mainly within the stem, but rarely in the root. We can assume this from the name that the Ancients gave it: they called this part of the plant the "heart" (*cor*) – others called it the "brain"[275] (*cerebrum*) or "matrix" (*matrix*) – with the idea that the principle of reproduction were derived from it. Because it is the central part, it is most fitting that it distributes life to the outer parts. To sum up, there are two distinct parts in plants, namely the root and everything that grows from bottom to top. The place which connects the root to the shoot [the root collar] seems to be a very appropriate location to find the heart of plants.[276]

14. [The plant's brain]
A certain substance appears at this place, sometimes emerging from the shoot, and sometimes from the root, which is softer and fleshier than these (hence being called "brain"), and is in many cases edible, as long as it is fresh: with time it becomes hard and woody, as do the other parts. The name "brain" seems most elegantly suited to this part. After all, in animals, brain marrow is found in the head, from which the marrow of the spinal cord originates and stretches out along its whole length. Similarly, in plants, the "brain" is situated in the root[277] as in a kind of head, which spreads the pith up throughout the length of the stem, a little like along a backbone, in order to distribute the vital liquid (*humor vitalis*) all the way to the furthest twigs and scions.

Chapter 2 [Nutrition]

15. [Animal nutrition]
Let us now look at how plants feed themselves and obtain their food. In animals, we can see that food is transported through the veins to the heart, as if to a workshop that produces inner heat. Once the food has been completely processed there, it is distributed by the arteries into the rest of the body by means of the breath (*spiritus*), itself

275 This is the usual translation of *cerebrum* from Pliny, *Historia Naturalis* (for example XIII, 9, §39–50) (ed. and Fr. transl. André et al. 1947–2015: 13.30–34; ed. and Engl. transl. Andrews et al. 1938–1962: 4.121–129).

276 This is also Aristotle's position in his *De Respiratione*. Although Aristotle does not assert that plants have a heart, they do have an equivalent part "between the middle of the growing stem and the root" which is "the part where the principle of existence can be found": XVII, 4, 478b33–479a1, our translation (ed. Ross 1955: 478–479).

277 This does not necessarily contradict the assertion of a scarcity of pith (and therefore "brain") in the root: it does indeed come from the root (or rather the root collar) and is distributed throughout the plant, but is therefore in a proportionately lower quantity in the root.

venas, neque alios ductus manifestos conspicimus, neque calorem vllum sentimus, vt mirum videatur, qua ratione in tantam magnitudinem arbores adolescant, cum longe minus calidi innati habere videantur quam animalia. <16> An animalibus ob sensum et motum, plurimum calidi innati datum est; et recta ratione minus adolescunt, quia multum nutrimenti absumitur in obeundis sensuum et motuum operationibus; plurimum enim eius conuertitur in spiritus: huius igitur gratia venas habent latas, vt mul-
[4] tum alimenti continere possint. <17> Plantae autem cum | soli officio nutritionis indulgeant, possunt pauciori calido innato, et magis adolescere et fructus multos edere. Quamuis autem sensui immanifestus sit calor, non ob id negandus est: quae enim minus calida sunt quam tactus noster, frigida indicantur. Venas quoque datas esse plantis licet exiguas argumento sunt illae, quae lacte manant vt Tithymalorum genus, et Ficus, nam si abscindantur, manat multus humor, vt incisa carne animalium sanguis: quod et in Vite maxime contingit, sed propter meatuum exiguitatem conspici nequaquam possunt; apparent autem in omni caule et radice quaedam vel instar neruorum fissilia secondum longitudinem, quos neruos vocant vt in Abiete: vel crassiora quaedam in ramos deducta, quas venas vocant in foliis plerisque manifestissima: hos igitur esse alimenti meatus putandum est, et proportione respondere cum venis animalium. <18> Non est autem in plantis vnus venarum caudex vt vena caua in animalibus, sed multae et tenues ex radice in cor, et ex corde in caulem ascendentes: non enim in ventre aliquo communi contineri opportuit alimentum, vt in corde animalium fuit necesse ad spirituum generationem: nam feruente humore multo simul in vase spiritus fit multus: sed satis fuit alterari humorem tactu medullae cordis, vt in animalibus cerebri medulla, aut iecoris caro facit: nam et in his non magnae et paucae, sed frequentes et

16 *hoc* impr.; C. corrigit. **23** *magnæ aut paruæ* impr.; C. corrigit.

produced in the heart by this same food. In plants, however, we find no veins nor any other visible channels, nor can we identify any heat. We might therefore wonder how trees are able to grow to such a height, whereas they seem to contain much less heat than animals.

16. [Relationship between nutrition and sensation and movement]
Perhaps a greater heat has been granted to animals because of their inherent capacity for feeling and moving. It would therefore be logical that they grow less, since they consume much food to meet the needs of their sensory and motor operations (*operatio*). Indeed, many of these foods are transformed into breath: this explains why animals have large veins, so that they can contain a plentiful supply of food.

17. [Heat and veins]
As for plants, as they | only have the single operation (*officium*) of nutrition, they can, [4]
with a lower internal temperature, both grow more and produce more fruit. However, although it is impossible to perceive their heat, it does not mean that they do not have any: for us, anything cooler than our skin feels cold. Furthermore, it is possible to accord veins to plants, based on the argument that they sometimes discharge a milky sap,[278] as is the case for plants in the euphorbia (*Tithymalus*) and fig tree (*Ficus*) genera. Indeed, if a notch is made in one of these plants, the liquid flows abundantly, just like blood does if an animal's flesh is pierced (this happens particularly in the case of the vine). Nevertheless, they [these veins] are so fine along their whole length as to be quite imperceptible. On the other hand, types of cracks, [which divide] like nerves, appear lengthwise in all stems and roots, and are in fact called "nerves", for instance in the fir. Sometimes they are thicker and extend as far as the twigs; these are [also] known as "veins" and are very visible in many leaves. We must consider them as pathways for the food and liken them to the veins in animals.

18. [Role of the veins and the pith of the heart]
However, plants do not have a single main vein from where the "veins" would come like the vena cava in animals, but instead a profusion of fine veins, which rise from the root to the heart and from the heart to the stem. Indeed, food is not contained in something comparable to the abdomen, which is necessary in animals to generate breath: [In them], breath only acts as effectively because much liquid is simmering at the same time in one container. But [in plants,] the liquid simply needs to be transformed by the action of the pith of the heart, just as that of the pith of the brain or the flesh of the liver does in animals. And in fact, in plants, it is not a case of a few large veins extending through it, but instead a large number of rather fine ones.

278 This milky sap is actually latex in the current botanical sense.

valde tenues sparguntur venae. <19> Fit autem per has alimenti attractio non motu aliquo fibrarum attrahentium, quod familiare est, et expellentium, quod fuerit contrarium: nam sine aliquo sensu alimenti haec non fiunt: huius enim gratia gustus et tactus datus est animalibus. <20> Cum igitur plantae omni sensi careant, genus alimenti non seligunt: sed humorem in terra mixtum alia ratione trahunt: quae autem ista fuerit, difficile est videre: neque enim ratione similitudinis dicere possumus, vt magnes ferrum trahit: nam in his, quod maius est, ad se trahit id, quod minus: quod si id fieret in plantis, cur potius humor terrae non traheret humorem plantarum? neque ratione vacui: cum enim in terra non solum humor contineatur: sed etiam aer, hic autem insequatur facilius vim attrahentis ob vacuum, quam humor, vt patet in cucurbitulis terrae admotis: plantae aere potius replerentur quam humore. <21> An quaedam sicca secundum naturam humorem trahunt? vt lintea, spongiae, pulueres: quaedam respuunt, vt auium quarundam plumae, Adiantum haerba, haec enim et si in aqua degant, non tamen madefiunt aqua ipsis non haerente: illa autem multum sorbent; quia cum

19. [Selection of food]
Food is obtained through these, not by some movement of fibers that would attract what is appropriate and would expel what is not: these functions do not operate in the absence of a sensation of food, hence why animals have been endowed with the senses of taste and touch.[279]

20. [Acquisition of food by plants]
Since plants have no sensation, they do not select their food, but instead draw their liquid, mixed with the soil, using some other means (*ratio*)[280] which is difficult to understand. Indeed, we cannot use the means (*ratio*) of a similitude to the way that a magnet attracts iron, because in the case of the magnet, the larger [object] attracts the smaller. If that were the case in plants, why would not the liquid in the soil, which is more powerful, attract the liquid from the plant? Neither is it by means (*ratio*) of a vacuum, since the soil contains not only liquid, but also air. However, the air is attracted by the vacuum with more force and less hindrance than the liquid is, just as in the case of cupping-glasses (*cucurbitula*) that you put close to the ground:[281] [in this hypothesis,] plants would be filled with more air than liquid.

21. [Moisture attraction]
Might this be a case of dry things which by nature attract moisture? This is the case of linen, sponges, and dust. But other things repel moisture, like the feathers of certain birds or the fronds (*haerba*) of southern maidenhair fern (*Adiantum*):[282] even when you soak them in water, they are not dampened, because the water does not attach to

---

279 Aristotle, in *De anima*, contrasts plants, which only have nutrition (*to threptikon*), with animals, which also have sensation (*to aisthètikon*). This is the condition for there to be pleasure and pain, and therefore the animal's faculty of desire (*to oretikon*) as the pursuit of the pleasant and the flight from the unpleasant. Animals at the very least have the sense of touch, which includes taste – the sensation of dry, wet, hot, and cold, and therefore of food. See 414a32–b10, 435b1–5 (ed. Ross 1961: 31 and 86). As for plants, they simply undergo the effects on their bodies without feeling them. See 424a30–b5 (ed. Ross 1961: 57).

280 By another "reason" in the old sense, meaning another logic, another process, another way of working. It is this *ratio* that Cesalpino is trying to convey in this reasoning.

281 In other words, when you apply a cupping-glass on the ground, if it were the case that a vacuum attracts water, then the cupping-glass would fill with water. But this is not the case: if the cupping-glass attracts anything, it is air; therefore, if the interior of plants were empty, it would draw air rather than water from the soil. See also Morton (1981a: 133) on this passage.

282 In the herbarium of Cesalpino, Adiantum refers to two species. The *Adiantum nigrum* (*Adiantum ruta-muraria*): the Wall-rue; the *Adiantum album* (*Adiantum capillus-veneris* L.): the Southern maidenhair fern also called Venus hair fern. The latter is known precisely for its hydrophobic character already reported by Theophrastus, *Historia plantarum* VII, 14, 1 (ed. and Fr. transl. Amigues 1988–2006: 4.37–38; ed. and Engl. transl. Hort 1916–1928: 2.135–137), Dioscorides, *De materia medica* 4, 134 (ed. Wellmann 1906–1914: 2.279–281; Engl. transl. Beck 2017: 239), and Pliny, *Historia naturalis* XXII, 30 (§62) (ed. and Fr. transl. André et al. 1947–2015: 22.43; ed. and Engl. transl. Andrews et al. 1938–1962: 6.207).

aqua conueniunt magis, quam cum aere. Ex huiusmodi igitur natura partes illas in plantis constare putandum est, quibus vtitur anima altrix ad trahendum alimentum: <22> Idcirco eae non ad venarum similitudinem meatu quodam continuo peruiae sunt: sed potius instar neruorum ex villosa constant substantia: sic enim bibula earum [5] natura continue humorem | ad principium caloris innati ducit, vt in lucernarum luminibus videmus; funiculo enim quodam vtuntur, quo oleum continue ad flammam ducatur. <23> Adiuuat autem hunc motum caliditas innata humorem affluentem absumens in germina et fructus: necesse est enim alium subinde consequi absumpto priori ob easdem causas, vt hi faciunt, qui penicillo in humore imposito, vt altera eius pars extra vas propendeat, humorem a feculentia secernunt: Distillante enim per penicillum exterius pendentem humore, alius continue subsequitur ascendens ex vase, donec vniuersus exeat clarus: sic radices plantarum ex terra, humorem puriorem continue bibunt, donec humor suppetat, et calor distribuat: quod enim ibi grauitas humoris facit, vt scilicet deorsum distillando prolabatur: hic leuitas praestat; calor enim hoc modo sursum ducit. <24> Ob id plantae pleraeque Vere et aestate germinant magis, et fructus edunt: quia a calore externo augetur humoris attractio: Hyeme latitat in profundo exilis flamma modico egens nutrimento, contra quam in animalibus natura calidioribus; viget enim in his calor innatus hyeme magis quam aestate. <25> Deficiente

them. [Among these things,] the former absorb a lot of water, because they are more compatible with water than with air. We must therefore consider that the parts of plants that the nutritive soul uses to obtain food are of a similar nature.

22. [Root absorption]
As a result, the roots do not travel through the plant in a continuous path as veins do, but just like nerves, consist of a tangled substance. Thus, their absorbent nature drives the liquid in a continuous way | towards what produces the internal heat, as we can see in the lights of lamps, which use a wick through which oil continuously supplies the flame.[283]

23. [Liquid filtration and purification]
However, it is the internal "warmth"[284] that makes this movement possible, by purifying the liquid that has flowed to the shoot and the fruit. Each drop can only follow the one that has flowed before through the same stems, in the same way that liquid is separated from its impurities by dipping a wick into it, so that one part of the liquid is drawn out of the container. As the liquid is filtered along the suspended wick, the rest continues to follow out of the container, emerging completely clear.[285] Likewise, the roots of plants continuously draw liquid from the soil, purifying it until it reaches the right place and is distributed by the heat. Indeed, the weight of the filtrate would risk dragging it down, whereas it rises if it is light enough: it is for this reason that the heat leads it upwards.

24. [Influence of the heat]
For this reason, many plants grow and fruit more in spring and in summer: procuring food is stimulated by the heat outside. In winter, the flame is weak, and it lacks even modest nourishment. It therefore hides out of reach, in contrast with what warmer animals' nature allows: their internal heat is more intense in the winter than in the summer.

283 In this chapter, Cesalpino tries to account for plant nutrition and selection of food by assimilating it with Aristotle's doctrine which asserts the absence of sensation in plants and therefore of desire and appetite. This attempt uses the model of physical movement, such as a body's fall, a magnet or liquid absorption, all of which are used as an analogy for appetite but are strictly determined by the nature of the body in question.

284 *Caliditas* (which is a later neologism), instead of the more usual *calor* that Cesalpino has used up to this point. Is Cesalpino making a distinction between heat itself and the state of being warm? If so, this may be a new way of referring to the principle of heat. This would be a way of breaking the circularity that his argument seems to introduce.

285 Here he is referring to capillary filtration: the water rises along the absorbent wick, leaving behind most of its impurities.

igitur aut diminuta humoris absumptione deficit, aut minuitur attractio: contingit autem id, vt summatim dicatur, vel defectu caloris insiti, vt in congelatione facta a frigore, et in aestu pabulum absumente citius, quam possit alter humor accedere: vtroque enim modo extinguitur calor innatus, quo extinto humor non digeritur; eo autem non distributo alius non attrahitur, vel ob uitium instrumenti, quo fit attractio. si enim hoc aut durius reddatur, quam oportet, vt in senectute fit, aut morbos alios patiatur: ad humoris transmissionem ineptum redditur: vnde plantas contabescere aut interire necesse est. <26> Quo igitur ingenio natura vtatur ad alimenti attractionem in plantis dictum est.

CAP. III.

<27> QVOMODO autem germinatio fiat, et cuius gratia, dicamus; sic enim caeterae quoque partes fient manifestae. Videtur autem propria esse plantarum germinatio, non enim in animalibus vllis huiusmodi affectionem perspicimus: illorum enim partes omnes factae sunt, antequam in lucem exeant. Plantae autem, quamdiu viuunt, nouellas edunt partes, quae germinatio vocatur. Forte autem in animalibus, pilorum, et dentium, et cornuum generatio germinationi similis videatur, cum posterius erumpant. Verum partes hae cum facultatem nullam habeant ab anima; sed vsum tantum praestent, merito non constituunt animalium substantiam: at plantarum germina tamquam animata vires eius animae et opera pulcherrima ostendunt. <28> solum in vtero geren-
[6] tibus videtur vera germinatio. Foetus enim, qui adnascitur, tam|quam germen eius partis viuit nutrimento ex eadem affluente. Differt autem; quoniam in his principium

25. [Distribution of the liquid]
From then on, inhibiting or decreasing the collection of liquid prevents or lessens the access to food. This can be explained, as we have touched on, either because the plant lacks internal heat (as in the case of freezing caused by the cold), and collects its sustenance more easily when it is warm, or because the liquid itself can or cannot be accessed. In either case, the internal heat is extinguished in the dried-out places where the liquid has not been distributed; the liquid did not reach these extinguished areas, for example because of a defect in the part of the plant responsible for drawing liquid. Indeed, if this part becomes harder (unavoidable especially in old age), or otherwise contracts diseases (*morbus*), it becomes incapable of transmitting liquid, with the effect that the plant inevitably weakens or dies.

26. [Conclusion on nutrition]
This is therefore the way (*quo ingenio*) that nature bestows on plants for obtaining their food.

Chapter 3 [Development and growth: embryo, bud, shoot and bark]

27. [Development]
Let us speak about how the budding process (*germinatio*)[286] takes place and what causes it: the other stages will then become clearer. Budding seems to be particular to plants, and we do not find the same phenomenon happening in animals: all parts of their bodies are formed before birth. Plants, on the other hand, generate new parts throughout their life, and we call this budding. It is true that in animals, the growth (*generatio*) of hair, teeth and horns seems to resemble this phenomenon, since they grow at a later stage. But in fact, these elements do not correspond to any faculties of the soul, but are connected only to practical functions, so are rightly not considered part of the substance of animals. However, the shoots (*germen*) of plants already contain the vital powers of the soul as well as its most beautiful operations.[287]

28. [Embryogenesis]
It is only in the uterus, in viviparous animals, that we can observe a true budding. The fetus, in fact, which "sprouts"[288] | like a sort of shoot (*germen*) in its own way, survives [6]

286 The term *germinatio* has no exact equivalent in contemporary terminology. It sometimes corresponds to what we would describe as a budding, sometimes to embryogenesis, sometimes to the development of the plant and sometimes to its growth or germination. See Greene 1983: 2.825–831; see also our commentary of chapters 3, 4, and 6.

287 Cesalpino will return to these essential operations in chapter 13: they refer to the functions of nutrition and growth on the one hand and reproduction on the other.

288 Here we have a play on the verb *adnascere*, which means both "to be born having been wished for" as well as "to grow" (as in teeth or plants).

extrinsecus ducitur, scilicet ex semine maris, alimentum ex vtero trahente; in plantis autem tum materia, tum principium intrinsecus prouenit: in Ouiparis tamen crescunt oua, et ea, quibus propendent sine semine maris, at infoecunda; sensualem enim animam, qua animal est, non habent sine maris coniunctione: hoc autem principio nequaquam eget ipsum vegetatiuum, quasi huius solius sit ex se germen emittere. <29> Ex his autem colligere illud quoque possumus: edendi fructus gratia germinationem datam esse; si enim in animalibus conceptus non fit sine aliqua germinatione, necessarium quoddam videatur ad eam propagationem, quae ex semine fit: sunt enim plantarum semina tamquam foetus animalium. Magis autem manifesta est in plantis germinatio, quia eae praeter radicem et cor, reliquas omnes partes ferendi fructus gratia habent: animalia autem solum vterum. Amplius quae gignuntur, in vtero latent ad partum vsque: quia foueri proprio calore foetum oportet: plantarum autem fructus, cum solis calore egeant ad perfectionem, non intro semper concludi oportuit: sed pro opportunitate explicari et ad lucem emergere fuit necessarium. Hoc autem facit ipsa germinatio; erumpit enim tamquam ex vtero omne germen et fructus, et paulatim in apertum exit: omnibus autem explicatis et in lucem editis, id totum, quod fructus gestat, videatur vterus inuersa, cui foetus multi sint appensi. <30> Ea autem pars, vnde germen erupturum est, oculus vocatur, quam solent accipere, qui inferendis arboribus student: turgente enim arbore vt cortex ob humorem subiectum facile abiungi possit a surculo, circumciso circa oculum cortice, auferunt, et alteri inferunt, in sedem eius plantae oculi iuste aptantes: quam inoculationem vocant; sic enim erumpit germen pro natura oculi insiti, extuberante autem iam germinis principio non dum autem explicato: partem hanc gemmam vocant, germinis autem explicatio fit dehiscentibus foliis: quanto enim magis erumpit germen, eo magis folia ipsum circumplectentia, tamquam manibus inuicem superpositis explicantur, solo pediculo adnexa: quasi huius gratia

thanks to the food that is provided, in the same way that a shoot does. There is a difference, however, in that in animals, the principle [of development] comes from the outside, that is by the semen of the male, although the food comes from the uterus. In plants, on the other hand, both the matter and the principle come from within[289]. Of course, in oviparous animals, the eggs grow, but those that are deprived of the male semen are infertile. Indeed, without the contact of the male, they do not have the sensitive soul by which animals are defined. As for the plant, it has no need for this principle; as if it is up to it alone to release the shoot from itself.

29. [Comparison of animal and plant embryogenesis]
To that, we can now add the following: development (*germinatio*) is devoted to the growth of the fruit. Indeed, if in animals conception does not take place without some sort of budding (which seems to be necessary for it to succeed), namely what the semen produces, then the seeds of plants are like the fetuses of animals. In animals, budding occurs only in the uterus, whereas in plants, it can be seen also in the root, the heart and all other parts that help bear the fruit. Animals are bigger from conception, and stay protected in the uterus up to the point of birth, because the fetus needs to be kept warm at the appropriate temperature. Conversely, the fruits of plants need the sun's warmth to be brought to perfection: they must not therefore remain enclosed, but should break free when the opportunity arises and emerge into the light. Budding itself works like this: the shoot and the fruit open out together as if from a womb, and gradually come out into the open air. Once all [the parts are] developed and conveyed to the light, the fruit-bearing whole resembles an inside-out uterus, from which several fetuses would be suspended.

30. [Bud and leaf development]
The place from which the shoot emerges is called the "eye" (*oculus*). Those who strive to graft (*inferre*) trees tend to detach this. Indeed, trees swell in such a way that the bark can easily be separated from the scion (*surculus*) because of the sap (*humor*) beneath it, and the bark peels off around the eye. Then [they] are separated and grafted onto another plant, and the eyes adapt precisely to the situation of that plant: this is called shield budding (*inoculatio*).[290] The growth of the inserted eye develops in this way, conforming to nature, after the principle of the shoot has swelled but before it has sprouted.[291] This part is called the bud (*gemma*), and the shoot unfurls at the same time as the leaves open: the more the shoot grows, the more the leaves spread around it; they unfold like hands superposed on each other, linked only by the peduncle (*pedi-*

289 Indeed, before R. J. Camerarius (1665–1721), the dominant idea was that plants do not reproduce sexually (unlike animals). See Camerarius 1694.
290 This refers to the technique of grafting from an eye or a bud rather than from a scion.
291 In other words, after the sap has risen in the spring to form the buds, but before they have opened and formed.

folia data sint, vt tenerum germen tueantur vel etiam fructum, vbi fructus erumpit cum germine: postquam vero eadem explicata sunt, alium vsum praestare videntur vmbram scilicet, ne a sole nimis vrantur tum fructus, tum nouella germina: moderatos enim solis radios vtraque desiderant, quod foliorum positione, et forma praestatur illos partim transmittentium, partim retinentium. <31> Ideo plurimis in autumno decidunt folia perfectis fructibus, et germinibus induratis. Si quae autem diutius seruant fructus, diutius quoque retinent folia: adeo vt quaedam vsque ad alteram germinationem et vlterius asseruent, vt Pinus, Arbutus, Laurus. Ferunt autem in regione feruente, vbi
[7] perpetui fe|re sunt aestus, nullis arboribus folia decidere: quod recta ratione fit: egent enim plantae perpetua inibi foliorum opera ad vmbram faciendam. Cum igitur folia ad tegendum data sint, merito ex cortice tantum ortum ducunt, quasi eius appendices quaedam, cortex enim conuestiendi gratia datus est. <32> Cum autem cortex ex duplici corpore constet, scilicet ex interno duriori robustiorique, qui in arboribus liber appellatur, et externo molliori quidem tenuiorique in iunioribus germinibus, in vetustis autem ob siccitatem asper redditur, et saepe dehiscit in arborum genere: folium in plerisque ex externi corticis substantia magis constat; vnde eius tenuitas et mollities oritur, et facile per siccitatem decidit secundum naturam quidem in autumno, praeter naturam autem in aestate. Quibus autem perpetuum est folium, his ex interni corticis substantia multum retineri in folio putandum est, sunt enim duriora et crassiora eorum folia, tam

culus).[292] It is for this reason, in a way, that plants have leaves: to protect the fragile shoot, as well as the fruit since it comes with it. After unfolding, they seem to fulfil another function, that of producing shade, so that the fruit does not burn from too much sun before new shoots appear. Both the shoot and the fruit need moderate rays, which the leaves provide by their position and shape, by letting some through and holding back others.

31. [Leaf conservation]

For this reason, leaves fall in the autumn: most fruit is ripe by then and the shoots have hardened. And if some [trees] retain their fruit for a long time, they will also keep their leaves for a long time, which is why certain species keep them until the following budding and even beyond, such as the pine, the strawberry tree (*Arbutus*) and the laurel (*Laurus*).[293] But in hot regions, where the heat is constant, | no tree loses its [7]
leaves! There is a good reason for this: in those regions, plants need the action of leaves to provide shade permanently. As the leaves are there as protection, it is only right that they grow so high from the bark as a sort of appendage to it, for the bark too is made to act as a covering.[294]

32. [Inner and outer bark]

Of course, the bark consists of a double body: inside there is a harder and more solid part that we call bast on the outside there is a layer which is softer and lighter in young shoots, but which is roughened in older shoots by the dryness [of the air], and which in the genus of trees is a cause of splitting. Nevertheless, the leaf tends rather to match the substance of the outer bark. This is the reason for its lightness and flexibility, as well as its deciduous nature stimulated by drought, conforming to nature in autumn, or prematurely (*praeter naturam*) in summer. For plants with evergreen leaves, we must assume that they have retained in their leaves more of the outer bark's substance, and their leaves moreover are harder and denser, as much in trees as in herbaceous plants. The veins, both those that run through the middle of the back[295] [of

292 The peduncle is the elongated and thin support of the organ of a plant (fruit, flower, leaf, etc.). In the current vocabulary of botany, the term peduncle is specifically used for the support of the fruit, and the term petiole for the support of the leaf. In order to remain faithful to the original text and not to necessitate more lexical precision, we have systematically translated *pediculus* everywhere by peduncle.

293 *De plantis* mentions two types of laurel: *Laurus sylvestris*, *laurustinus* (or *laurustinus tinus*: *Viburnum tinus* L.) and *Laurus regia* (cherry laurel: *Prunus laurocerasus* L.). It is likely that when the epithet is omitted, as it is here, Cesalpino is referring to the laurustinus. Indeed, the cherry laurel was only discovered by Pierre Belon (1517–1564) in the Empire of Trebizond in 1546 (before being imported into Italy and then into the rest of Europe).

294 Here Cesalpino anticipates somewhat his later argument (§32 and following). He argues that the leaves come from the bark (and not from the wood or the pith) of which they are an extension and thus share the protective nature.

295 The term *dorsum* refers to the back of a mount or beast of burden, where the load is placed.

in arboribus, quam in humili materia. Venae autem, tum quae per medium dorsum, tum quae in latera sparguntur, ex interiori cortice oriuntur; hic enim solus venosus est; quod autem ex cortice constent folia, illud argumento est, quod in quibusdam videmus totum corticem in folia transire, vt nullus relinquatur germen conuestiens praeter folium: vt in Harundine, Tritico, Ferula: sed in his magis folij pediculus quam folium caulem circum amplectitur substantia crassior existens, quam folium. <33> Cuius igitur gratia data sunt folia germinantibus, et qui sit eorum ortus, ex dictis contemplari licet. <34> Germinis autem substantia interius principium ducit; non enim sine medulla est, neque sine eo corpore, quod medullae circumponitur; si quis enim virgam germinantem denudare a cortice studeat, auferet quidem cum cortice folia, vtpote quae illi soli sint appensa, germina autem non auferet; continua enim sunt cum interni corporis substantia: idque recta ratione; Cum enim germen ferendi fructus gratia datus sit, vis autem prolifica in medulla tamquam in corde contineatur: hanc per tota germina deductam esse oportuit: quod si hanc, et lignum quoque circumpositum medullae, aut aliud corpus huiusmodi et extremum corticem. <35> Videntur autem quaedam sine medulla viuere, vt Calami, et quae caulem concauum ferunt: sunt et arbores, quarum caudices excauati viuunt, vt Salix, Olea. An medullae natura in his non seiuncta est a circumposito corpore seu ligno, et forte in nullis? oportet enim venas alimentum trahentes tangi a medulla, eae autem per totam caudicis substantiam feruntur: quantum igitur medullae in medio est venas non tangens, crescente caule, aut caudice euanescit, locum concauum reliquens: in nouellis enim germinibus, seu a radice, seu ex ramis nulla reperiuntur concaua. <36> Dubitabit quoque quis non in medulla; sed in cortice magis vim prolificam esse: inoculatio enim fit solo cortice insito; germinat autem se-
[8] cundum naturam corticis, non secundum naturam | ligni subiecti. Praeterea pleraeque

the leaf] as well as those on the sides, come from the inner bark: only this layer is venous. But since the leaves match the outer bark, one argument follows that we can see in the leaves proof that the whole of the bark extends into them, so that nothing remains of the shoot except the leaf that covers it, as is the case in reeds (*Harundo*), wheat (*Triticum*) and plants in the *giant fennel* (*Ferula*). However, in these species it is the leaf peduncle more than the leaf itself that surrounds the stem, as it is made of a tougher substance than the leaf.

33. [Function of the leaf]
From what we have stated already, we can therefore see why developing plants are endowed with leaves, and where these leaves come from.

34. [The shoot]
The shoot's substance is linked to an even deeper principle. Indeed, the shoot does not exist without the pith nor without its outer body. If we strive to strip a rod (*virga*) of its bark, we also remove the leaves along with the bark (since they are only appendages of it), but we do not remove the shoots, because they are connected to the substance of the inner body. And this is for a reason: since the purpose of the shoot is to bear fruit, and the prolific force is contained as much in the pith as in the heart, it is fitting that this force feeds both the shoots in their entirety, the wood (or any analogous body) that surrounds the pith as well as the outermost bark.

35. [Disappearance of the pith]
Some plants with a hollow stem, such as rush (*Calamus*), seem for that matter to exist without pith. The trunks of certain trees survive despite being hollow, such as the willow (*Salix*) and the olive tree. Is the nature of the pith in these plants not separated from the surrounding body, in other words the wood? Would the same not be true of all species? It is fitting that the veins that bring the sustenance are in contact with the pith. Furthermore, they spread across the whole substance of the trunk. The pith that is not in contact with the veins therefore disappears when the stem or trunk grows, leaving a hollow area.[296] Indeed, we do not find any hollow areas in the young shoots, nor in the root, nor in the twigs.

36. [Vital role of the bark]
We might wonder if the prolific force might not be found in the bark rather than in the pith, since grafting is done only on the bark. The graft produces a bud in accordance with the nature of the bark, not of the wood | beneath it. Furthermore, many trees die [8]

296 Cesalpino means that the fact that certain trunks are hollow is because it is the veins that bring the sustenance, and therefore only the pith that surrounds them can or must stay alive; if the veins follow the shape of the bark and therefore are spaced apart, the inner pith is not in contact with them and therefore disappears.

arbores circumciso in orbem solo cortice emoriuntur in totum, non emoriuntur autem excauata medulla. <37> Dicimus corticem insitum ideo germinare; quia prorumpit ex subiecto ligno germen, cui cortex agglutinatur ob affinitatem; nisi enim corticis oculus, oculo ligni accomodetur, non germinat; producuntur autem et folia et fructus secundum naturam corticis: quia haec in omnibus ex cortice ortum ducunt: semina autem interna non secundum naturam corticis, sed ligni subiecti fiunt; nam si serantur, nascuntur, non pro natura insiti corticis, sed vt plurimum syluestre genus: ortum enim ducunt ex medulla, non cortice; quod autem corticis in orbem circumcisio in plerisque arborem enecet, fit; quia germinatio non fit sine cortice, vt superius est explicatum, qui autem derelinquitur supra circumcisum locum, emoritur; quia ablatus est alimenti ductus ex infernis; si quis autem ex ligno subiecto ali possit, viuit, vt contigit Suberi, et aliis quibusdam, quibus pertinaciter cortex ligno inhaeret.

CAP. IIII.

<38> Prima autem plantarum germinatio est a radice, quod enim primo nascitur siue ex semine, siue per putredinem radix est. Cum autem in eius capite sit ea pars, quae cordi animalium respondet, inde germinis principium ducitur, aliquando vno caule, aliquando pluribus; cor enim in quibusdam indiuiduum est, vnde vnicus caulis assurgit, vt in plerisque arboribus: in quibusdam quodam modo diuisionem patitur, vnde ab eadem radice multi caules erumpunt, vt in Tritico. <39> Numquam autem vnum et idem cor author est plurium caulium: patet autem in iis, quae amisso priori caule alterum fundunt, vt in ferulaceo genere; numquam enim ex eadem parte, sed a latere germinant; Vt autem vnum cor vnius est caulis, sic videretur vnius quoque radicis

---

16 *inde*] *vnde* impr.; C. corrigit.

completely if one makes a notch in the bark all the way around, whereas they do not die if their pith is hollowed out.[297]

37. [Role of the bark in growth]

For this reason, we say that the grafted bark buds, because the shoot emerges on the outside of the wood beneath it, to which the bark is connected. But if the eye of the bark does not match the eye of the wood, there will be no budding. In contrast, the leaves and the fruit are produced by the nature of the bark, because their growth is generated from the bark. On the other hand, the seeds inside [the fruit] are not made from the nature of the bark, but from the wood beneath it. Indeed, if they are sown, they appear, not as the same nature as the implanted bark, but rather like the wood: their growth comes from the pith and not from the bark. Conversely, what in many cases makes a tree die when it is notched all the way around depends on the bark. Indeed, budding cannot take place without bark (as we have explained above), and what remains above the cut point dies, as it is separated from its nutritive source, which comes from below. Nevertheless, if something can be fed from the wood beneath the bark, then it survives, as in the case of the cork oak (*Suber*) as well as in other species whose bark is firmly attached to the wood beneath.

Chapter 4 [Growth and development: heart, stem, soboles and leaves]

38. [Heart of plants]

The first stage of a plant's development happens from the root, as this is what is born first – either from the seed, or from putrefaction.[298] But since the part which corresponds to the heart in animals can be found at its top, the principle of growth is linked to it, sometimes by a single stem, sometimes by several. Indeed, in certain plants, the heart is undivided, thus explaining why only one stem grows, as is the case in most trees. In others [plants], a kind of division takes place, which explains the growth of several [hearts] from one single root, as in wheat.

39. [Heart–stem–root relationship]

In contrast, one single heart is never at the origin of several stems. It can nevertheless happen that plants that have lost their former stem generate another one, as in the ferulaceous genus. Furthermore, this new stem never grows exactly in the same place [as the previous one], but instead next to it. Since the same heart is linked only to a

297 This observation was already made by Theophrastus, *Historia plantarum* IV, 15, 1; IV, 16, 4, (ed. Amigues 1988–2006: 2.114–117; ed. and Engl. transl. Hort 1916–1928: 1.405 and 104, respectively), and *De causis plantarum* V, 17, 1 (ed. and Fr. transl. Amigues 2012–2017: 3.51–52; ed. and Engl. transl. Einarson and Link 1976–1990: 3.181–183).

298 That is, spontaneous generation.

vnum cor esse, quod si hoc esset, nullae essent, quae plures caules ferrent. An radix vna vnum quoque germen initio profert; postquam autem eadem magnitudinem adepta est conspicuam, diuiditur in quibusdam in plura principia vel simul, vt in Tritico, vel successiue, vt vno extincto alterum subnascatur vt in Ferula. <40> Plantarum enim plurimarum natura est, vt diuisae viuant: quia earum principium licet actu vnum sit, est tamen potentia plura: nihil autem refert, siue a nobis plantae diuisio fiat, siue sponte diuidatur principium solum: diuisae enim radices quaedam, licet in paruas partes concisae sint, germinant vt Graminis, Raphani Montani: nam vbique erat cor potentia, et haec radice seri possunt: sponte autem multitudo germinum fit, vel eadem subiacente radice communi, vt in Foeniculo: vel singulis germinum principiis noua
[9] suborta radicula, vt in Cipero, | Iride et tandem geniculatis fere omnibus: quasi sponte natura ipsa radicum diuisionem in geniculis moliatur. <41> Inter has quaedam sunt, in quibus noua orta radice et germine, altera tamquam senio confecta emoritur, vt in Satyrio, et Gladiolo contingit; diuidit autem natura ex toto cum germinum principiis etiam nouas radiculas, in Allio et caeteris Bulbaceis: Allij enim spicae appellatae principia sunt germinum omnia ab eadem radice orta, qua exsiccata propriae singulis suboriuntur radiculae, qui propagandi modus per sobolem appellatur. Germinatio igitur ex radice hoc modo habet; exoritur autem in hac germinatione caulis quidem in humili materia, caudex autem in arborea; <42> cum autem omne germen fructus ferendi

single stem, it would seem that the same heart is also only linked to one root; however, if that were the case, no plant would bear several stems.[299] Besides, the same root also only generates one shoot at the beginning, but once this has grown enough to be visible, it divides in some plants into several principles, either at the same time (as in wheat), or successively, so that once one dies, another is born (as in giant fennel).

40. [Division of plants]
Indeed, the nature of many plants is such that they live divided, because it is possible that their principle is single in its act, but multiple in its potentiality. There is no reason why this should not be the case, either the plant is divided by us, or the principle divides itself. Indeed, certain roots, once they are divided, can subdivide into smaller parts and develop, as in the case of scutch grass (*Gramen*) or horseradish (*Raphanus Montanus*). Here, the heart is a potentiality, and [these parts] can be propagated by the root: thus a multitude of shoots develops naturally, either all together around the same root – like fennel (*Foeniculum*) –, or in the form of young cormlets (*radiculum*) develop-
ing from the particular principle of the shoots – like chufa sedge (*Ciperus*), | iris (*Iris*) [9]
and almost all Geniculeous plants (*Geniculatum*). The very nature of the roots stimulates division in the joints of the nodes (*geniculi*), so to speak from themselves (*quasi sponte*).[300]

41. [Reproduction by soboles]
Among these plants, some form a new root and a new shoot, while others die, weakened by age, as happens with the snake's head iris (*Satyrium)* and the gladiolus (*Gladiolus*).[301] On the other hand, in allium and other bulbaceous plants (*Bulbaceum*), nature creates more new young cormlets through division by means of the principles of shoots in their entirety: the principles of garlic shoots are known as cloves (*spica*). They all originate from the same root, from which individual young cormlets come one after the other, depleting it of its energy. This is known as propagation by means of soboles (*per sobolem*). This method of development comes from the root. The stem (in herbaceous plants) or the trunk (in woody plants) draws its origin from this process.

299 This argument is quite difficult to grasp but is as follows: the root is at the origin of the heart (the root collar) and of its potential division. The heart is at the origin of the growth of one single stem. So if a plant has several stems, it needs several hearts which have been generated by the multiplication of the root. A plant that has multiple stems is therefore, by necessity, linked to several roots via its different hearts.

300 Cesalpino does not have a clear-cut usage of this word which he employs to describe different situations at different points in the text.

301 The plant that we call gladiolus today is, however, a bulbous plant, contrary to what the text implies.

gratia datum sit, quod a radice est, aut statim fructus gestat, vt in Bulbaceis, et frumentorum generibus, et in Palma, aut aliis intercedentibus germinibus, idque vel per vnum vel per plura media, secunda igitur germinatio in caule fit ramos constituens, tertia in ramis ramusculos producens: et sic deinceps. appellant autem extremas germinationes in arboribus surculos, quos ad inserendum assumunt. In genere igitur humili, quae caulem ammittunt, nullae fere transeunt tertiam germinationem: in genere autem arboreo, ad minus tertia germinatio fructificat, vt in Vite, idque non ex semine, sed si seratur ramo, singulis annis nouo germine prodeunte, non omnibus vno anno erumpentibus, vt in genere herbaceo. Germinis autem eruptio, non ex omni parte caulis fit: sed maiori ex parte ex alis foliorum, qua scilicet parte folij pediculus nectitur cauli, inibi enim oculus est futuri germinis, quasi folium eius custodiendi gratia sit appositum, sinu quodam in cortice relicto, qua ab eodem exoriens abscedit. Fit enim in hac parte veluti alterum cor erumpente in eam sedem ex interna medulla principio. <43> Quapropter nodus quidam caulis apparet, qui si totum caulem cingat, geniculus vocatur, vt in Tritico et Calamo, coeuntibus et veluti implicatis in hac parte neruis, partim ad robur, in quibus caulis est inanis, vt in Tritico, partim ad noui germinis productionem, quibus germinatio in caule data est. <44> Ordo autem quidam germinum spectatur vt foliorum, nam vel singula in singulis nodis, siue geniculis nascuntur, idque vel in duobus lateribus tantum, haec quidem in dextris, haec vero in sinistris alternatim, vt in Arundine, vel pluribus, in quibus ordo minus perspicitur, vel in singulis geniculis bina, aut etiam plura ferunt folia, ac germina; quae autem bina ferunt, alternatim disposita sunt, vt haec quidem ante et retro spectent, haec vero in dextra et sinistra: ob quam foliorum et germinum positionem in pluribus caulis redditur quadratus, vt in Marrubio. <45> Horum igitur omnium a latere germinatio fit; sic enim disposita sunt folia, ex quorum alis erumpunt germina; vnde et rami alae vocantur;

42. [Stem ramification]

But since any shoot is made to bear fruit, which it does from the root, either it carries the fruit directly (as in the case of Bulbaceous plants, the genus of Cereals (*Frumentum*) and the date palm (*Palma*), or via one or more other shoots: a second stem development produces branches (*ramus*), and a tertiary one, from the branches, produces twigs (*ramusculus*), and so on. In trees, these shoots are known as "scions" (*surculus*), and are used for grafting. In herbaceous plants, plants which lose their stem go through a third phase of having no stem. Returning to trees, only the third stage of the ramification process produces fruit: in the case of the vine, for example, if it is planted from a twig rather than a seed, a new shoot is produced each year, even if all the branching stages do not happen in the same year, unlike in herbaceous plants. Furthermore, the development of the shoot does not take place entirely from the stem, but predominantly from the stipules (*alum*) of the leaf, in other words where the leaf peduncle is attached to the stem. At this place, the eye coincides with the place of the future shoot; it is attached to it in order to protect it as a leaf would do, forming a sort of hollow in the bark from where it emerges. What is more, it is in this part of the plant that a kind of second heart is operating, appearing by the action (*principium*) of the inner pith.

43. [Node]

For this reason,[302] a sort of node appears on the stem, and if it encircles the entire stem, it is known as a leaf sheath (*geniculus*), as in the case of wheat and rush. Fibers (*nervus*)[303] are linked to it and as if intertwined, sometimes with the tough part (*robur*) in the case of plants with a hollow stem (like wheat), sometimes [with the inner portion] to produce a new shoot in the case of plants where the development takes place within the stem.

44. [Leaf arrangement]

A series of shoots then appears, which look like leaves. They emerge sometimes one per node or leaf sheath, alternately on each side, one on the right and the other on the left each time, like on reeds, sometimes several per node, in a less ordered arrangement. In the case of the latter, we can observe either one pair per node, or more leaves or shoots. The leaves which emerge in pairs are arranged alternately; we can see either one in front and the other behind, or one on the right and the other on the left. As a result, the square well represents the position of the leaves and shoots in many stems, as in the case of the white horehound (*Marrubium*).

45. [Stipules]

Budding of these stems takes place laterally; indeed, this is the arrangement of the leaves which grow from the buds of the stipules (*alum*) from where the shoots emerge.

302 In other words, to protect this new heart.

303 In Latin and in Greek, this term is used both for an animal nerve in a body and for a plant fiber.

[10] sparguntur enim a | latere instar alarum: <46> Quaedam autem sunt, quarum germinatio non in omni foliorum exortu; sed in summo tantum fit veluti fisso in ea parte caule, vt Tithymalorum genus, Linum, Lithospermon, et inter arbores Abies, Pinus: omnibus enim his folia vndique dense vestiunt caules, vt non pateat germinis exortus nisi in summo: ideo in his rami ex certo interuallo et numero conditi pulchre spectantur, quod in caeteris non videtur: licet enim geniculi, vel nodi certo interuallo distent, non tamen in omnibus germina proueniunt, vel non paria. <47> Germinationes igitur ordine dispositae a radice vsque ad extremos surculos fructus ferentes huiusmodi sunt.

CAP. V.

<48> QVONIAM vero plantarum principium, quod cor appellatur, non quemadmodum in animalibus in vno quodam loco seiunctum permanet: sed etiam veluti in omnes partes distributum est, fit, vt multae diuisae non solum viuant, vt quaedam animalia inter insecta faciunt: sed per diuisionem propagentur, quod nullis animalibus contingit: vbicunque enim est cordis natura, ibidem est et germinis et radicis principium: ante diuisionem quidem potentia, postea vero etiam actu, idcirco radices quaedam abscissae germen emittunt, et e contra caules quidam abscissi, si intra terram condantur, radices egerunt, et alimentum trahunt: quamquam non similis in omnibus sit propensio: quaedam enim radice diuisa melius proueniunt, vt Helenium, Raphanus montanus, <49> quaedam caudice, aut ramis maioribus seri postulant, vt Olea, quaedam surculis, vt Punica et multa in genere suffruticum: Magis autem radices ex vetustioribus ramis erumpunt: germina autem ex iunioribus: idcirco, qui serunt, surculo aliquid annotini rami relinquunt, vt in Vitium malleolis. <50> Quod autem ob iniuriam diuisio-

11 *sed etiam veluti*] *sed veluti* impr.; C. corrigit. **18–19** *vt Helenium. Raphanus montanus,* impr.; corrigimus. **20** *suffruticum*] *suffructitum* impr.; corrigimus

It is for this reason that we call them "wings":[304] they unfold | from each side like wings. [10]

46. [Apical growth]

The budding of some plants, however, does not come entirely from the leaves, but takes place in a sort of fissure right at the top of the stem,[305] as in the genus of euphorbias, flaxes (*Linum*), gromwells (*Lithospermum*) and, in trees, firs and pines. In all these plants, the leaves densely cover all parts of the stems, so that the development (*exortus*) of the shoot is only possible right at the top. This is why we can observe branches placed harmoniously according to a certain number of determined intervals, a harmony which does not occur in other plants, because the joints of the stems (*geniculus*) either extend their nodes according to determined intervals, but do not produce shoots at all of these places, or [they do not produce] equally spaced ones.

47. [Conclusion on development]

The development processes are thus arranged in an orderly fashion from the root up to the furthest ramifications (*surculus*) which bear the fruit.

Chapter 5 [Vegetative reproduction]

48. [Propagation by division]

Since in reality the principle of plants, that which we call the "heart", does not reside in a certain place of its own, as it does in animals, but can be found, as it were, diffused through all their parts, it follows that most of them can not only survive when they are divided, as do, among animals, some insects, but also propagate by division, which is not the case of any animal. Alongside the nature of the heart, lies also the principle of the shoot and of the root, potential before division, and actual once it has taken place. This explains why certain roots, when cut, sprout shoots and conversely why certain stems, when cut and planted in the soil, produce roots [which] feed them. However, this tendency does not happen in the same way in all species: certain plants reproduce even better when their root is divided, like elecampane (*Helenium*) and horseradish.

49. [Cuttings]

Some [trees] need to be planted by the trunk or by the largest branches, like the olive tree, others by scions, like the pomegranate and the majority of undershrubs (*suffrutex*). More roots issue from older branches, but more shoots from younger ones. That is why those who plant them leave a portion of the branch attached to the scion, as in the layering (*malleolus*) of vines.

---

304 In Latin, the word for "stipule" is *alum*, which also means "wing."

305 Cesalpino is probably describing apical growth.

nis multae faciunt, idem sine diuisione quaedam moliuntur: nam si contingat ramos terram attingere, radices in contactu dimittunt intra terram, et nouam prolem germinant, quas viuiradices vocant, vt in Vite et fere omnibus, quae ramo, aut surculo proueniunt. Maxime autem id fit in herbaceis quibusdam, quorum caules humi serpunt, vt in Hedera terrestri, Ranuculo quodam, Pentafillo, et Humirubo: nam in his, duo genera caulium natura moliri videtur, vnum, qui attollitur ad fructum ferendum: alterum qui serpit ad sobolem in singulis geniculis procreandam: idem quaedam intra terram faciunt, vt Gramen, et Ciperus; nam et hae in singulis geniculis et radicant, et germen emittunt. <51> Illud autem peculiare adnotauimus in minori Dentaria: cum enim haec
[11] caulem rectum ferat; radiculas tamen geniculatas in singulis foliorum alis fert, quibus | decidentibus in terram noua erumpit planta, cum tamen in summis caulibus eadem siliquas ferat, in quibus semina continentur, vt simul in eodem caule vtrumque moliatur semen et sobolem. Quamuis autem alias sint, quarum caules radices edant supra terram, non tamen ad sobolem gigniendam: sed auiditate nutrimenti illas egerunt, vt Hedera, quae suo amplexu arbores, vel parietes numerosis radiculis exugit, et Melica, ac Semperuiui quoddam genus arborescens: hae namque radiculas ex caule longas dimittunt, donec terram attingant. <52> Peculiare quoque in Allio et Porro videatur: cum enim haec sobolem ad radices ferant, gigniunt etiam in summis caulibus loco seminis. Quae enim seruntur, radicum capita sunt non semina, similem sobolem etiam

50. [Layering and suckering]

Whereas many plants are damaged when they are divided, certain become weaker if they are not divided. Indeed, if branches reach the ground, the roots [created] upon contact spread into the soil and produce buds on a new rootstock: we call this an offset plant (*viviradix*), as in the case of the vine and almost all species which sprout from a branch or scion. This happens most often in certain herbaceous plants, whose stems creep on the ground, like ground-ivy (*Hedera terrestris*), a certain ranunculus (*quidam Ranunculus*),[306] creeping cinquefoil (*Pentafillus*)[307] and the dewberry (*Humirubus*).[308] In these species, nature seems to have established two types of stem: one that rises to bear fruit, the other that scrambles to produce a sobole at each joint. Some do the same underground, like the scutch grass or chufa sedge: these too push out roots and shoots at each joint.

51. [Soboles/bulbils]

We have noticed this particularly in the case of the coral root (*Dentaria minor*):[309] while it grows a straight stem, it bears little articulated roots[310] (*geniculus*) at each of the stipules of its leaves. When they | fall, a new plant takes root in the soil, even [11]
though [the coral root] itself bears its seed-containing siliques (*siliqua*) at the top of its stems, as if it had put into its stem both the seeds and the sobole. Although there are other plants whose stems produce roots above ground level, in those cases it is not to generate soboles, but for the desire of nourishment, like the ivy (*Hedera*), which in its grip sucks trees or walls by its numerous rootlets, and melic grass (*Melica*), or types of evergreen trees, whose long radicles propagate from the stem, until they touch the ground.[311]

52. [Bulbils]

There is also something particular that we observe in garlic and leek (*Porrum*):[312] while their soboles are found next to the roots, it is actually at the top of the stems that the seed is produced. Indeed, when we sow them, it is not the tips (*caput*) of the roots

306 Probably creeping buttercup: *Ranunculus repens* L.

307 It is creeping tormentil (*Potentilla reptans* L.) whose other name, cinquefoil, better reflects the original Latin name.

308 This plant is not mentioned in Cesalpino's herbarium but it is described in book 14, ch. 18. For the identification of the dewberry (*Rubus caesius* L.), see Dodoens 1554: 6.475–476.

309 Cesalpino is likely to be referring here to *Cardamine bulbifera* (L.) Crantz in today's nomenclature. We call it coral-wort or coralroot bittercress. It used to be recorded under the scientific name *Dentaria bulbifera*.

310 *Radicula geniculata*: he is probably referring to the bulbils here.

311 Cesalpino is probably not talking about conifers here, but rather plants like figs, which do indeed produce aerial roots. He is possibly referring to tropical (evergreen) plants such as the strangler fig or mangrove trees.

312 The description that follows corresponds to the perennial leek (*Allium ampeloprasum* L.) and perennial garlic or sand leek (rocambole) (*Allium scorodoprasum* L.).

in summis foliorum vidimus in quodam Bulbo, quem Moli vocant: <53> Differt autem soboles a semine, vt foetus viuens ab ouo: semen enim tamquam ouum est, in quo est principium vitale; at vita nequaquam. soboles autem viuit primo quidem iuxta parentem, vt eius germen. postea vero per se ipsam propriis radicibus ex terra humorem trahens. Praeterea soboles, vel radix est incoata, vel germen, vel vtrumque conspicuam magnitudinem habens: semen eorum principium gerit inclusum in cortice: idcirco nutrire multam sobolem plantae nequeunt, semina autem valde numerosa ferre possunt, vt in animalibus Viuipera et Ouipera se habent: sobolis autem generatio simplicior est tamquam ex particulae auulsione propagatio. Seminis constiutio multiplices partes requirit. <54> Postquam igitur et radicum, et germinum ortum et constitutionem explicauimus, deinceps seminum apparatur aggrediamur.

CAP. VI.

<55> Cvm in ea propagatione, quae fit ex semine, plantarum finis consistat, quae enim ex sobole fit imperfectionem naturam sequitur, qua scilicet plantae diuisae viuunt, merito plantarum pulchritudo in seminis productione maxime ostenditur: Nam et numero partium, et figuris, et conceptaculorum differentiis fructificatio longe maiorem ornatum praesefert quam germinatio: adde florum admirabilem venustatem, qui genitricis naturae delitias in seminibus condendis praemonstrant. <56> Quemadmo-

---

8 *Viuipera*] *Vipera* impr.; C. corrigit.

that act as the seed, but rather what we observe above the leaves that looks like a sobole in a sort of bulb (*Bulbus*), which we call a bulbil (*Moles*).

53. [Differences between the sobole and the seed]
The sobole differs from the seed as the fetus does from the egg: the seed is like an egg which contains a principle of life, but not a life in itself, whereas the sobole has a life of its own; albeit near and like a shoot from its parent, before it is able to draw its liquid from the soil by itself with its own roots. Previously, the sobole, when it is big enough to be visible, is either an incomplete root, or a shoot, or both. The principle of these [also] produces the seed within the bark.[313] As a result, plants are not able to nourish too many offspring, whereas they have no problem in bearing many seeds, as can viviparous and oviparous animals. On the other hand, the generation of the sobole is simpler, since it consists of a release, while the formation of seed requires several stages.

54. [Conclusion on roots and shoots]
Now that we have explained the origin and the formation of roots and shoots, let us turn to the properties of seeds.

Chapter 6 [Seeds]

55. [Importance of seed production]
The raison d'être of plants is reproduction (*propagatio*)[314] from their seed (furthermore, the imperfect nature of plants – surviving even when divided – arises from [the possibility of propagating] the sobole). It is therefore with good reason that the beauty of plants manifests itself in the production of seed;[315] Indeed, fructification (*fructificatio*) is much better endowed than growth (*germinatio*), in regard to its number of parts, shapes (*figura*) and varieties of the receptacles (*conceptaculum*); to this you can add the wonderful charm of the flowers, which give a foretaste of Mother Nature's delights in the seeds that must develop.

313 Since the sobole is being contrasted with the seed, he must be referring to the seed of the shoot and the root. But the seed that is produced is of course found on the outside, even if it is being generated from within.
314 A more usual translation would be "layering" but that would be inappropriate here.
315 This argument rests on the Platonic limited/unlimited opposition (in other words, determinate/indeterminate) which is often superimposed on the Aristotelian form/matter compound: the limit or the form is what makes something stable and gives it its beauty, whereas the unlimited or the matter is a lack of form, in other words an imperfection.

dum igitur in animalibus semen excrementum est vltimi alimenti in corde, cuius calore vitali cum spiritu apprenhenso foecundum redditur: sic in plantis necesse est seminum substantiam ex ea parte secedere, in qua principium est caloris innati, quam medullam esse superius comprobauimus. Hinc igitur ex humidiori puriorique parte alimenti seminis medulla exoritur, ex crassiori autem eiusdem putamina ad tutelam
[12] circumposita. <57> Non fuit autem necesse in plantis genituram ali|quam distinctam a materia secerni, vt in animalibus, quae mare et foemina distinguuntur: maris enim genitura materiam in foemina efformat sua vtens corpulentia tamquam instrumento, vt in Quaestionibus Peripateticis explicauimus. At plantae cum non multa egeant organorum distinctione, pauciorique spiritu vtantur, materiam simul cum spiritu formante secernunt, qui conceptus est tamquam ouum: idcirco maris et foeminae distinctione non indiguerunt, quamquam secundum aliquam similitudinem quaedam mares appellentur, quaedam foeminae. <58> Inest igitur in omni semine quaedam plantae incoatio. Quemadmodum enim in ouo quaedam particula continetur, in qua est animalis futuri veluti delineatio, reliquum autem corpulentiae pro alimento est: sic in plantarum seminibus pars illa principatum continet, vnde radix erumpit et germen; est enim quasi corculum quoddam reliqua parte seminis alimentum illi primum subministrante: patet autem id maxime in Tritico; pars enim quaedam inest veluti oculus, quam si laceras, nequaquam nascitur: ob id eam particulam formicae erodere solent, antequam recondant intra terram. Si autem reliqua pars seminis vulneretur, aut a vermicu-

7 *animalibus*] *animalibns* impr.; corrigimus

56. [Origin of the seed]

Just as the seed in animals is taken from the completely refined (*ultimus*) food in the heart, whose life-giving warmth and breath make it fertile, in plants the substance of seeds comes from the part containing the principle of internal warmth, which we have shown to be above the pith[316]. On this basis, the pith of the seed comes from the moist and pure portion of the food, whereas the external layers, which surround the pith in order to protect it, come from the coarser portion of the food.

57. [Non-sexual seed production]

On the other hand, in plants, we have not needed to consider[317] that the task of generation | is carried out by anything other than matter, in contrast to animals which are [12] divided into males and females: the sperm of the male (*genitura*) gives form to matter in the female by using her corporeality (*corpulentia*), as we explained in *Quaestiones peripateticae*.[318] But as plants do not need a great differentiation of their organs and use less breath, they organize their matter thanks to a formative breath conceived as in the case of an egg.[319] This is why they are not lacking the male/female distinction, even if we do name them as male and female in accordance with some comparison.

58. [Seed anatomy]

In each seed is situated a kind of beginning of the plant. Indeed, just as an egg contains a sort of small part with a sketch of the animal-to-be, while the rest of the contents (*corpulentia*) of the egg functions as its food, similarly in the seeds of plants, this part contains the principle from which the root and the shoot arise. There is therefore a sort of little heart[320] while the rest of the seed serves as its first food; this is particularly apparent in wheat. Indeed, a certain part similar to an eye is found there: if you damage it, nothing will be born. It is for this reason that ants often remove this small section, before burrowing underground again[321]. If the remaining part of the seed is damaged or infested with worms, nothing at all will emerge from the heart either, even if that is left intact.

316 Cesalpino is alluding to chapter 3 (§34–35) where he showed that the principle of life in a plant is distinct from the pith and can be found just above it, even though it gives life to it.

317 Cesalpino is referring to the argument that he develops in chapter 3 (§28–29).

318 Cesalpino 1571: V, 2: the male brings only the form, whereas the female contributes both matter and form, and so it is the corporeality of the female that the male exploits.

319 As he shows later, Cesalpino is comparing the seed (which does not need fertilization) to a laid egg that is *already fertilized*: the seed development does not depend on external fertilization, at least from a male principle (see commentary for more details).

320 This is the embryo.

321 To prevent the seeds that the ants have collected from germinating in their anthill, they remove the embryo before storing the seeds underground.

lis perforetur, integro seruato corde nihilominus prouenit. <59> Est autem seminis medulla in plurimis candida, sicut et animalium genitura: principia enim corporum naturalium vt plurimum candida sunt, vt elementa, siquae tamen flaua sunt vt multorum leguminum semina, ad albedinem aliquo pacto vergunt. <60> Pinguedinem praetera plurima praeseferunt: adeo vt ex quibusdam oleum extrahatur, vt ex Nucibus, Amigdalis, Sinapi: Humidum enim vitale, in quo spiritus continetur, calidum est et aereum, non aqueum: Non tamen humida sunt, vt animalium genitura; longe enim siccior est plantarum natura quam animalium; at molliciem cum ea siccitate quandam habent, cum praecipue fuerint humectata: nam circa exortum omnia humidiora sunt, per aetates autem continue exsiccantur. Quamuis autem quaedam semina ossea ex toto videantur, vt Palmae, vel cartilaginea, vt Rusci, Gladioli, et aliorum quorundam: latet tamen seminis medulla mollior in exiguo quodam meatu, vnde erumpit in exortu plantula integro osse derelicto. <61> Sicca igitur cum propria mollicie semina esse oportet, cum perfecta sunt, exempta scilicet a sole omni aquea humiditate non aerea: sic enim sata humorem facilius trahunt e terra, quo ante omnia intumescunt: Deinde excitato ignis principio in ipsis latente, vt Calci contingit, in humoris occursu, idem humor cum lactea seminis substantia permixtus et concoctus, tamquam familiare alimentum auget conceptum antea incoatum; Tunc autem radix primo emergit peciolo quodam ex corde seminis prodeunte, qua corticem dehiscere et egressum semini concedere necesse est: postquam autem radicem in terram egerit, reliqua seminis corpu-

59. [Seed color]

The pith of the seed is white in most cases, like the sperm of animals: the principles of life in natural bodies are usually white, like germs (*elementum*),[322] and all those which are yellow, like the seeds of many legumes (*legumen*), tend somehow to have a whitish hue.

60. [Seed consistency]

In addition, many [seeds] contain fat, as in the case of those from which we extract oil, such as walnuts (*Nux*), almonds (*Amigdala*) and mustard (*Sinapi*). Indeed, life-giving moisture (*humidum vitale*),[323] in which breath is contained, is warmth and air (*aereus*),[324] and not water. They are not wet (*humida*)[325] like animal sperm, because the nature of plants is much drier than that of animals, and a certain softness accompanies this dryness in them, even when they are watered. Furthermore, when they emerge from the soil, all plants are damper, but they progressively dry over the years. Of course, some seeds seem to be entirely "bony", such as those of the date palm, or "cartilaginous", such as those of the butcher's broom (*Ruscus*), gladiolus and some others. However, the softer seed pith nestles in a narrow passage, from which the seedling grows when it emerges from the ground, leaving behind it all its stone (*os*).[326]

61. [Seed germination]

It is therefore fitting that seeds are dry and sufficiently soft when they are at maturation, in other words drained by the sun of all their non-aerial watery moisture. Indeed, that way the plants, when sown, can more easily draw liquid from the soil from where they grow before anything else. Then, once the principle of fire[327] that they harbor is activated on contact with liquid – as happens with limestone (*Calx*)[328] –, this liquid is mixed and prepared with the milky substance in the seed, just like the nutrients which feed a fetus before birth, to take a more common example. At that point, the root emerges from a first petiole (*peciolum*),[329] as if it came from the heart of the seed, just where the bark must open and leave a passageway for the seed. Once the root is fixed

322 *Elementum* is generally translated by "element." However, this translation makes little sense here, so we chose the secondary sense of the term which designates the beginning of something, so here, the germ.

323 Cesalpino must be referring to the vital fluid which he usually refers to as *humor*.

324 Usually the term *aereus* means bronze or copper, but it can also be a rare variant of *aerius*, air, which makes much more sense here.

325 The proximity between *humor* (vital liquid), *humidum* (vital moisture) and *humida* (wet) is difficult to render in English.

326 The word *os* is used both for the stone and any material that is rigid or like bone.

327 Cesalpino is presumably still talking about inner heat here, although he uses the word *ignis* here rather than *calor* or *caliditas*.

328 Cesalpino refers here to quick lime, calcium oxide, which reacts violently with water producing heat, according to the reaction: $CaO + H_2O \rightarrow Ca(OH)$.

329 This would be a radicle in modern terminology.

[13] lentia in plurimis ex suo cortice | tamquam ex ouo in lucem prodit, quae in duo foliola pulposa explicata particulam ostendit, vnde germen erupturum est: Qua enim haec duo folia exortum ducunt, cor est, quippe radicis caput, et germinis principium, sunt autem haec alterius generis folia, quam quae in germinatione exoriuntur: illa enim tantum ad tutelam data sunt tenuia ex solo cortice orta, haec partes sunt seminis ad alimentum primum cordi ministrandum: ideo crassa existunt; tota enim ferme seminum pulpa in his consistit: <62> ob id magna pars seminum in duas partes dissecta est corde tantum coniungente; si quae autem sunt, in quibus diuisio nulla sit, vt in grano Tritici: seminis pulpa in exortu minime egreditur e cortice, germen vero e latere erumpit integra seminis corpulentia cordi haerente, idem quoque fit in multis leguminibus, quamuis eorum semina bifida sint: cum enim germen ex latere erumpat non ex medio, seminis partes non explicantur in folia, ob id neque exeunt e cortice: manent autem iuxta nouellas plantas huiusmodi seminis appendices donec totus humor lacteus cum sua pinguedine in partium nutrimentum transierit, quo tempore exsiccatae decidunt. <63> De ortu igitur seminum et natura, ac substantia dictum est, de ceteris autem partibus seminis gratia datis deinceps dicamus.

CAP. VII.

<64> QVEMADMODVM in animalibus circumuolui foetum membranis oportuit ad tutelam, sic seminibus plantarum multiplicia data sunt inuoluctra ob eandem causam: sed

in the soil, the remainder of the seed's body (*corpulentia*) usually emerges into the light, out of its bark | like an egg. The unfurling of two pulpy leaves[330] indicates the point at which the shoot will emerge. Indeed, at the point where these two leaves appear is also where the heart is [located], as well as the head of the root and the principle of the shoot. But these [two] leaves are of a different kind to those that develop during growth: the latter are made specifically delicate enough to protect [the fruit][331] and come only from the bark; the former are parts of the seed whose function is to provide the first nourishment from the heart. This is why they are thick: almost all seeds contain pulp (*pulpa*) within them.[332] [13]

62. [Seed division]

As a result, a large number of seeds are divided into two parts linked by the heart (although there are some without any division), as in a grain of wheat. The seed's pulp protrudes very slightly from the bark at its origin, but the shoot emerges from the side, and the whole body (*corpulentia*) of the seed remains attached to the heart; this even happens in many legumes, certainly those with bifid seeds. When the shoot emerges from the side and not from the middle, the parts of the seed do not develop into leaves, and because of this are not released from the coat, like appendages to the seed, but instead remain next to the new seedlings, until all the milky fluid has been transferred with its oil as nourishment for the organs (*partes*), at which point they detach, dried over time.[333]

63. [Conclusion on seeds]

We have dealt with the origin of seeds, their nature, and their substance; now let us turn to the other parts whose purpose is to work with the seed.

Chapter 7 [Flowers]

64. [Different seed husks]

Just as the fetus is suitably surrounded by membranes for its protection, so the seeds of plants are endowed with many husks (*involucrum*)[334] for the same purpose. Some of these, for example the flowers, fall even before the seed has finished its maturation;

330 He is referring to what we would now call cotyledons.

331 If the leaves must be slender, it is to protect the fruit by letting through only a proportion of the sun's rays, as Cesalpino explains in chapter 3: §30–31.

332 This pulp which gives a thick consistency to the seed and the emerging cotyledons is in fact the seed's endosperm as we would term it today.

333 This passage is quite difficult to understand and can generate two interpretations, see our commentary.

334 These seed husks do not necessarily correspond to the current botanical term which includes the tegmen and the testa of the seeds.

eorum quaedam abscedunt incoato semine nondum perfecto, vt flores, quaedam postquam semen absolutum est, vt pleraquae vascula et siliquae; dehiscunt enim sponte exsiccato semine; quaedam perpetuo haerent quousque semen germinare incipiat; tunc enim intumescente medulla cortex abrumpitur. <65> Flores igitur partim ex necessitate, partim ad tuendos fructus incipientes dati sunt; ex necessitate quidem, quoniam turgente planta, vt in Venere solent animalia, efflari necesse est aliquem spiritum; non enim sine spiritu fit seminis eruptio: quod autem ex huiusmodi substantia flores orti sint, manifestum est: id enim indicat et substantiae tenuitas, ex qua constant, et odores, quos vt plurimum spirant: <66> indicat et mellea dulcedo, quae in plerisque reperitur, quamque apes seligentes in suos alueolos recondunt: vt enim mel aereum ex percocta a sole exhalatione nascitur roris modo ex aere decidens, sic ex plantae halitu percocto, qua parte egressum habet, veluti sublimatum, floris concamerationi et staminibus haeret, alterum genus mellis, quod apes colligunt, et fauis repo-
[14] nunt. <67> Quod autem interim ad tegendos | fructus flores dati sint, patet: priusquam enim explicentur, aut insident ipsis fructibus, vt in Rosa, Malo, Piro, aut illos vndique

4 *abrumpitur*] *absrumpitur* impr.; corrigimus

others [fall] after the seed has been released, such as the majority of capsules (*vasculum*)[335] and siliques (indeed, these detach on their own when the seed dries out); others linger on the plant until the beginning of the seed's development (indeed, at this moment, the bark[336] breaks as the pith expands).[337]

65. [Functional and mechanical causes of flowers]
Flowers therefore partly exist to protect the fruits in their early stages, and partly out of necessity). Indeed, when the plant swells, as animals often do during their mating season, it must exhale something. It is a fact that the seed cannot be issued without breath. The flowers are manifestly made from a substance that is a little like this breath. Moreover, the lightness of the substance of which they are made is an indication of this, as well as the scents that they emanate in most cases[338].

66. [Parallel with honey production]
The sweetness of honey, which is often present when bees store it in their alveoli after selective harvesting, indicates the same. Indeed, copper-colored honey is produced by condensation from the air, in the way that dew is produced by steam warmed by the sun. In the same way, another kind of honey (from the one that bees harvest and deposit in their alveoli) remains attached by the filaments (*stamen*) and the corolla (*concameratio*) of the flower;[339] it is as sublimated by the steam warmed by the plant, at the point at which it leaves it.[340]

67. [Protective function of flowers]
Flowers evidently exist | to protect the fruit while they are on the plant: indeed, before [14]
they unfold, either they take the place of the fruit itself – as in the case of the rose (*Rosa*), the apple tree (*Malus*) or the pear tree (*Pirus*) –, or they surround it completely

---

335 The term *vasculum* has the same meaning in Pliny as capsules in their current sense: *Historia naturalis* XV, 28 (§115) and XVIII, 7 (§52) (ed. and Fr. transl. André et al. 1947–2015: 15.57 and 18.75). However, in chapters 9 and 10 of this book, Cesalpino uses the term in a wider sense, to include anything that contains a seed, such as a segment of a citrus fruit or the pericarp of a fig.

336 The term *cortex* refers to what botanists today would call spermoderm or integument and is what Cesalpino will later subdivide into the internal layer (which corresponds to the integument) and the external layer (which corresponds to the testa).

337 He is referring here to the seed's germination, which punctures the husks.

338 Cesalpino is showing firstly that there must exist some sort of breath (*aliquem spiritum*) which makes the issue of seed possible. He then argues that flowers must be made from something light and fragrant, which is also a kind of breath (the breath of the flower). He is thus associating two types of breath.

339 He is most likely referring to the nectar of flowers.

340 The process that he describes here is the reverse of sublimation in chemistry, since he is looking at a transformation from the gaseous state of breath to the solid state of pollen, in other words a sort of condensation.

circumplectuntur, vt in Amigdala, Pruno, Olea: crescente autem fructu, iam flores aperiuntur, ac paulo post tamquam in futurum inutiles decidunt arefacti. cum enim non ex qualibet materia nutriantur: sed alituosa, quae una cum semine erumpit, necesse est ea deficiente marcescere, cum praesertim a sole citissime huiusmodi substantia resoluatur. Quem igitur usum folia tenellis germinibus praestant, eumdem flores nascentibus fructibus opitulantur; plurimi enim foliati sunt. <68> Differunt autem, quia folia explicata alium vsum praebent, quem superius diximus: Flores explicati omnino inutiles redduntur. retinentur igitur folia diuturno tempore, et propter vsum: et quia continue affluit nutrimentum ex cortice, flores cito decidunt ob causas contrarias; <69> color quoque foliorum herbaceus est: humor enim plantarum soli expositus hunc colorem contrahit, vt in aquis stagnantibus apparet, cum exsiccantur a sole: Flores autem nulli, aut paucissimi herbacei sunt, omnino virides certe nulli: cum tamen caeteri colores omnes excepto nigro in floribus spectentur, iique venustissimi, modo simplices, modo omni varietate commixti quasi natura de industria pingente. hoc enim est spirituum sublimatorum opus, vt in subterraneis apparet, nam tingente exhalatione et lapides et pulueres omni colorum varietate spectantur; natura enim spiritus ob materiae puritatem sinceros recipit colores, ob tenuitatem vero eosdem facile commu-

– like the almond tree, the plum tree (*Prunus*) or the olive tree.[341] When the fruit grows, the flowers cease at that point to cover it; a little later, they fade and fall, [becoming] in a way useless for the future. Moreover, the flowers are not fed by just anything, but by the nourishing [substance] that comes together with the seed; when that is no longer present, withering is inevitable, and especially when the substance leaves [the seed] in this way because of the great intensity of the sun.[342] The usefulness of the leaves for the tender shoots is the same as that of the flowers for the growing fruits; what is more, many of them are endowed with leaves.[343]

68. [Different durability of leaves and flowers]
Flowers, however, differ from leaves because the latter, once they are unfurled, provide another function that we have already mentioned; as for flowers, once they are open, they become completely useless.[344] The leaves are therefore preserved for a long time because of their usefulness and because food continues to flow into them from the bark; for the opposite reason, the flowers fall quickly.

69. [Color]
Furthermore, the color of the leaves is the same as grass, because the liquid of plants, once it is exposed to the sun, takes on this color, just as we can see in stagnant water, when it is dried up by the sun.[345] On the other hand, no flowers, or very few, resemble grasses, and in any case they are not at all green; on the contrary, we see all sorts of other colors in flowers, except black, and they are all very pleasant – sometimes solid colors, sometimes mixed in all different ways, as if painted industriously by nature. It is in fact the work of sublime breath, for by the action of damp exhalations as we can see underground, stones and dust of the greatest variety of colors present themselves: the nature of breath receives, thanks to the purity of the material, base pigments that can be easily mixed together because of their fine grain. From here come as much the

341 It is difficult to understand the botanical contrast that the author is making here. He might be referring to the distinction between fruits that come from an inferior ovary on the one hand and those from a superior ovary on the other, in terms of their position in relation to the parts of the flower, but the examples of species that he gives do not correspond with this distinction. See our commentary.
342 Literally: by the very close sun.
343 The idea is therefore that the flower fulfils the same function to the fruit as the leaf does to the shoot: it protects it from the sun. These "leaves" (*folia*) of the flower could refer just as well to the sepals as to actual leaves, or even to the petals (see the commentary for more details).
344 In chapter 3 (§31), Cesalpino writes that certain plants keep their leaves longer because they continue to need shade (or a filtering of the sun's rays).
345 Here Cesalpino is describing the green deposit, which in fact is made up of green (micro)algae which is left when stagnant water evaporates. It is therefore not the water itself that turns into a green liquid.

tabiles: Vnde vel puncta vel tenuissimae lineae versicolores oriuntur. Merito autem virides nulli visuntur flores, quia eorum materia non humor est, sed potius spiritus: neque nigri ex toto; nigredo enim aut humorem sequitur semiustum, aut cuiusque corruptionem; neutrum autem in florum generatione inest: <70> constant autem flores plerique ex folio, et stamine et floccis; folium in circuitu magis ambit, modo vnicum et indiuisum, concauum, vel patulum, vt in Smilace, Anchusa, modo in plura folia dissectum, vt in Papauere, Rosa, stamina in medio tantum foliorum sunt ex summa parte fructus egredientia, quippe quae ex sede seminum vlterius producta nascantur, vt in Croco et Lilio terna; quia seminum sedes in vasculo tripartita est: Flocci in ambitu

markings as the very fine lines in shimmering colors.[346] But no flower appears green, for the good reason that its substance is not a liquid, but a more powerful breath. They are not black at all either, since blackness either comes from half-consumed liquid,[347] or from its decay, but neither of these is found in the production of flowers.

70. [Morphology of flowers]

On the contrary, most flowers are made up of leaves [petals],[348] a filament (*stamen*)[349] and flakes (*floccus*).[350] The leaf [the petal] roughly forms a circle, sometimes uniformly and in one piece, concave or open, like in hedge bindweed (*Smilax*) and viper's bugloss (*Anchusa*[351]),[352] sometimes dividing into several leaves [petals], as in the case of the common poppy (*Papaver*)[353] and the rose.[354] The filaments are so well centered within the leaves [petals] that they rise above the greater part of the fruit, because they have been produced such that they have been pushed from the point of the implantation (*sedes*) of the seeds, which is triple, as in the crocus (*Crocum*) and the lily (*Lilium*),[355]

346 Cesalpino could be saying that just as the underground breaths paint different colors, in the same way the breath of plants paints colors on its flowers. He could also be suggesting that the pictorial action of these breaths is the cause of the colors in the flowers, since the plants draw their food from the soil.

347 The liquid being referred to here is most likely the vital and nourishing liquid that we have seen extracted from the soil, and that constitutes the matter of leaves; this apparently does not stop it from being able to be burned.

348 He might also be referring in a non-specific way to the perianth, that is all the sepals that form the calyx and all the petals that form the corolla, but a floral anatomical vocabulary had not been completed at that time. The term "petal" only appears for the first time in 1649 in the work of Fabio Colonna (see the commentary).

349 The term *stamen* literally means a thread, which has led to the modern term stamen. However, in Cesalpino's writings, the term could describe both the stamens and the pistils of a flower: see the commentary for discussion.

350 He is most likely referring to the anthers which sit on top of the filaments of the stamens and contain the pollen.

351 *Anchusa* on its own without its adjectives can refer in *De plantis* to three different species, all of which belong to the family of plants today known as *Boraginaceae*. The first two are viper's buglosses and belong to the same genus: Italian viper's bugloss (*Echium italicum* L.) and purple viper's bugloss (*Echium plantagineum* L.). The third is the common gromwell (*Buglossoides arvensis* (L.) I. M. Johnst.).

352 Bindweed, and to a lesser extent viper's bugloss (and generally all plants in the *Boraginaceae* group) are gamopetalous (or sympetalous) plants, which is to say that their corolla is made up of a single circular or concave petal (or pseudo petals, more or less split, but fused at the base).

353 In *De plantis, Papaver* can refer either to the common poppy (*Papaver rhoeas* L.) or the blackspot horn poppy (*Glaucium corniculatum* (L.) Rudolph).

354 These flowers are dialypetalous, in other words their corolla is formed from several separate petals.

355 Cesalpino uses the noun *Lilium* for three different species in *De plantis*: lily of the valley (*Convallaria majalis* L.); orange lily (*Lilium bulbiferum* L.); martagon lily (*Lilium martagon* L.).

magis sunt iuxta folia, pendent autem corpuscula quaedam ex tenuissimis filamentis, quorum ortus similis videtur fungis, qui in lucernarum luminibus ex fuligine oriuntur: hinc enim multum spirant flores, numerosa autem sunt huiusmodi corpuscula, in quibus semina numerosa sunt, vt in Papauere, quasi haec sint singulorum seminum propagines: patet id in Chamaemelo, et Ranunculo: nam praeter folia ambientia singulis seminibus proprij insident flocci potius quam flosculi. <71> Non sunt autem in
[15] omnibus manifesta vel folia, vel flocci, vel stamina propter exiguitatem: qua|propter aliquando his carere putantur: in quibusdam folia adeo tenuia sunt, vt flocci videantur lanuginei, vt in Thaliethro, muscosos flores huiusmodi vocant. <72> Transeunt in quibusdam flores in aliam substantiam, fructus ipsos praeter lapsi, vt in Nuce Auellana, Castanea, et tandem Amentiferis omnibus. Amentum enim pro flore est corpus

whose implantation of the seeds is tripartite within the capsule.[356] The flakes [anthers] are closer together in the circle than the leaves [petals] are, and are attached like small bodies to very fine filaments, whose origin looks like the mushrooms that comes from the soot in lamp openings.[357] Furthermore, many flowers breathe (*spirant*)[358] through these, and such little bodies are numerous, containing numerous "seeds",[359] as in the poppy, they are like suckers (*propagines*)[360] of the individual seeds. This is obvious in chamomile (*Chamaemelon*) and buttercups: before their leaves [petals] unfurl in a circle, flakes implant themselves on individual seeds more firmly than the little flowers (*flosculus*)[361] do.

71. [All parts are not necessarily visible]

However, because of the small size of certain plants, the leaves [petals], flakes or
filaments are not always visible. This | is why one sometimes thinks that they are [15]
missing; in certain plants, the leaves [petals] are so fine that the flakes seem fluffy, as in meadow-rue (*Thaliethrum*);[362] we call these mossy flowers (*muscosos flores*).

72. [Catkins as flowers that have become fruit]

In some plants, the flowers, before they fall, turn into another substance, that is the fruits themselves, as is the case of the hazel (*Nux Avellana*), the sweet chestnut (*Casta-*

---

356 The ovary of the crocus does indeed have a three-celled capsule. We must understand from this that the housing of the lily and crocus seeds is a triple one, since the housing of the seeds is tripartite within the receptacle in these species. From today's technical perspective, the ovary has three chambers which contain three ovules, which will develop into three seeds contained in a fruit that we call a trilocular capsule.

357 What we call a mushroom wick soot is the result of poor combustion caused by impurities in the fuel, leading to a sooty-black accumulation of residue at the tip of a candle wick or in the opening of a lamp. The analogy is therefore a technical one (see illustration in the commentary).

358 According to Aristotle and in the Aristotelian tradition, plants do not breathe. See *De anima*, 410b25–411a1 (ed. Ross 1961: 22–23). Is Cesalpino straying from this tradition here, or is he simply using this term in an analogical sense (which would explain the use of "but" that follows: plants breathe like animals do, but using more numerous organs)? It is also possible that the author simply means that the flowers exhale their fragrance from the anthers.

359 As he clarifies just after this, these "seeds" are not true seeds, but rather their "offshoots"; he is probably referring to the grains of pollen contained in the anthers.

360 The stamens are "rooted" on the ovaries – the receptacle that contains the seeds; they are fed by and therefore unfold from the fruit (the ovary and its seeds), and their anther thus contains the by-products, like a new plant in miniature.

361 The term *flosculus* does not seem to refer to a particular organ in a technical sense. One might imagine that this could be a reference to the capitulum (or floret) of the chamomile, for example, if it were not for the buttercup not having such a flower composed of capitula. The term most likely therefore refers to a small flower, the size of that of the chamomile or buttercup.

362 This must undoubtedly be identified as Siberian columbine meadow-rue, also known as French meadow-rue (*Thalictrum aquilegiifolium* L.) in which the downy inflorescences are made up of numerous stamens, but no petals (see illustration in the commentary).

oblongum, ex sede fructus productum: ideo fructus sine flore spectantur manifesto; stamina enim in Amenti longitudinem transeunt, folia autem et flocci in eius squamas degenerant. simile quid videatur quod in Milij Indici summis calamis effunditur spicarum modo; Flores enim, qui singulis seminum paniculis distribui debuissent, in huiusmodi Amenti genus abierunt. <73> Nec absimilis fuerit Ricini flos; nam et hic sine fructu est, non multum tamen illum praeter lapsus; quod et in herbis quibusdam notauimus. sunt etiam quaedam, in quibus Amentaceum quid oritur sine vlla spe fructus; steriles enim omnino sunt; Quae autem fructum ferunt, non florent, vt Oxycedrus, Taxus, et in genere herbaceo, Mercurialis, Vrtica, Cannabis; quorum omnium steriles mares vocant, foeminas autem fructiferas: Quod ideo fieri videtur; quia foeminae materia temperatior sit, maris autem calidior; quod enim in fructum transire debuisset, ob superfluam caliditatem euanuit in flores, in eo tamen genere foeminas melius prouenire et foecundiores fieri aiunt, si iuxta mares serantur, vt in Palma est animaduersum, quasi halitus quidam ex mare efflans debilem foeminae calorem expleat ad

nea) and all the amentiferous plants (*Amentiferum*).[363] A catkin (*Amentum*) is an oblong body which replaces the flower; the fruit is produced from its implantation. This is why the fruit appears without an obvious flower. Indeed, the filaments transform into the long part of the catkin, while the leaves [petals] and flakes degenerate into scales (*squamas*). Something similar appears at the top of the sticks of sorghum (*Milium Indicum*), distributed like thorns.[364] Indeed, the flowers, which should have been situated within each panicle of seeds, are transformed into catkins.

73. [Sterile plants, male and female]

The flowers of the castor bean (*Ricinus*) are not so different: before they fall, they are fruitless and sparse.[365] Besides, certain [plants] are completely sterile, including amentaceous plants (*Amentaceum*), which are born without any hope of producing fruit. As for [sterile plants] which do bear fruit, they do not flower, such as cade (*Oxycedrus*), yew (*Taxus*) and in the herbs genus (*herba*), mercury (*Mercurialis*), nettle (*Urtica*)[366] and hemp (*Cannabis*).[367] In all these species, sterile plants are called male, and those that fruit female. This is for the reason that the female plant's matter is more tepid and the male's hotter; because what should have passed into the fruit vanishes in flowering due to the excessive heat.[368] However, it is said that in this type [of plant], females grow better and are more fertile if they are planted near males. It is noticeable in the case of the date palm, for example, that a sort of breath emitted from the male plant compensates for the lack of heat in the female for fructification.[369]

---

363 This is used for plants with flower heads in strings or catkins (walnut, hazel, hornbeam, etc.).

364 In the "sticks" (i.e., panicles) of *Milium Indicum* (sorghum, great millet, *Sorghum bicolor* (L.) Moench), the racemes are indeed finely distributed like thorns. The term referring to the species, already used by Pliny, has a long history. See Franconie 1997.

365 Cesalpino probably means that the flowers of the castor bean (i.e., the male flowers) fall before the fruit (i.e., the female flowers) has developed.

366 *De plantis* only describes the Roman nettle (*Urtica pilulifera* L.) (Book IV, ch. 45). However, this species is monoecious, unlike the common nettle (*Urtica dioica* L.) which is dioecious, and which is probably being referred to here.

367 Cesalpino is dealing here with dioecious plants, i.e., male plants that have inflorescences that do not fruit, and female plants that bear fruit but whose inflorescences do not necessarily look like flowers.

368 The excessive heat is entirely transferred to the flowers, and there is not enough left to make the fruit.

369 We have known since Prospero Alpini (1553–1617) that there is a fertilization that takes place in date palms, demonstrated scientifically in 1580. However, the practice is even older and was described by Theophrastus *Historia plantarum* II, 8 (ed. and Fr. transl. Amigues 1988–2006: 1.64–66; ed. and Engl. transl. Hort 1916–1928: 1.151–153); *De causis plantarum* II, 9 (ed. and Fr. transl. Amigues 2012–2017: 1.89; ed. and Engl. transl. Einarson and Link 1976–1990: 1.279–281), and then passed on in less precise detail by Pliny *Naturalis historia* XIII, 7 (§29–35) (ed. and Fr. transl. André et al. 1947–2015: 13.27–29; ed. and Engl. transl. Andrews et al. 1938–1962: 4.115–119). Strictly speaking these observations do not lead to the recognition of sexuality in plants by Cesalpino, because as we can see here, it is rather a type of breath that leads to fruit generation.

fructificandum. <74> Omnis vero flos cum ex intimis partibus, vnde seminis materia erumpit, ortum ducat, exterius tegitur altero inuolucro, qui calyx vocatur in quibusdam vt Rosa et Papauere: oritur autem hic cum extimus sit ex caulis cortice, seu tunica, ideo herbacei coloris est, vt folia, dehiscit autem vna cum flore; sed non decidit vt flos; nutritur enim ex cortice, et saepe eo vtitur natura ad fructum continendum, vt in Ormino et Acanaceis: semina enim eorum in calyce floris continentur, postquam flos deciderit: peculiare in Vesicaria et Cicorio, quod eorum calyces dehiscunt quidem, dum flores panduntur: sed postquam deciderint, contrahuntur vndique fructum amplectentes, vt Cicorij semina eximi nequeant, nisi madefiant; sic enim conceptacula aperiuntur. Haec vel continuo constant corpore, vt in Polemonia, Calamento: excepta ea parte, qua flos erumpit, vel in plures partes dissecta, vt in Rosa, Cicorio, Acanaceis, Frumentis; sunt enim in plerisque veluti squamae inuicem amplexae. <75> Haeret autem huiusmodi floris tegmen in quibusdam adeo pertinaciter, vt disiungi nequeat a flore, vt in Ornithogalo; flos enim eius interius lacteus est, exterius herbaceus ob caulis tunicam: idem Elleboro contingit, quod in causa est, vt vniuersus flos quodammodo herbaceus sit, et non decidat; retinetur enim a tunica ediuncta: Haeret quoque similiter et
[16] ipsis fructibus, in quibus flos aut | nullus est manifestus, vt in Ficu, aut in summo fructu insidet, vt in Cucurbita, Punica; continuus enim est in his floris calyx, cum extimo fructus cortice: quibus autem huiusmodi cortex abiunctus est a fructu, flos in

74. [Flower calyx]

In truth, every flower, while it originates from its inner parts from where the seed matter emerges, is covered on the outside by another husk, called a calyx (*calyx*) in certain plants, like in the rose and the poppy. But the origin of the calyx is right on the outside of the stem's bark, in other words its tunic (*tunica*); that is why it is the same color as grass, like the leaves, and it opens together with the flower, but it does not fall like the flower does. Indeed, it is fed by the bark, and often nature uses it to contain the fruit, like in the case of wild asparagus (*Orminum*)[370] and of Acanaceous plants: their seeds are contained in the flower's calyx after the flower has fallen. The balloon vine (*Vesicaria*) and chicory (*Cicorium*) have the peculiarity of unfurling their calyx only as long as the flowers are open, but after they have fallen, they close up entirely to envelop the fruit, with the result that chicory seeds cannot be removed if they are not wet, because that is the only way that these receptacles will open. The calyxes either consist of a body that is continuous, as in the case of bladder campion (*Polemonia*) and catnip (*Calamentus*) – except for the place where the flowers emerge – or divide themselves into several parts, like in the case of the rose, chicory, plants in the Acanaceous and cereals, most of which have types of scales which wrap around each other.

75. [Unification of the fruit and flower and its protection]

In this way, in certain [plants], what protects (*tegmen*) the flower is fixed so firmly that one cannot separate it, as in the ornithogalum (*Ornithogalum*): its flower is white (*lacteus*) inside, but grassy (*herbaceus*) on the outside because of the stem's tunic.[371] We can see the same in the hellebore (*Elleborus*), which explains why the whole flower is, in a way, herbaceous and does not fall: it is held by the tunic that is attached to it.[372] It also holds onto its fruits in the same way, either by clinging onto them directly, in the case of | those in which no flower appears, like the fig tree, or by hanging onto the tip [16]
of the fruit, like the marrow (*Cucurbita*) or the pomegranate. Indeed, in these [species], the calyx of the flower is continuous; the fruit is continuous with the outermost bark. However, in these, the bark is separated from the fruit such that the flower's origin is inside the root of the fruit: every flower is contained within the bark of the stem.

370 Cesalpino associates the Greek (*h*)*orminum* with wild asparagus (V, 26: 218; see also XI, 15: 441 and XI, 29: 448–449). The word, however, can also mean sage, which would be a better example of what he is describing. The calyx in sage does indeed contain the fruit, but not that of asparagus. Based on Cesalpino's herbarium, Caruel (1853: 115) identifies *Horminum sylvestre* as bastard balm (*Melittis melissophyllum* L.).

371 Ornithogalum is a plant that has tepals, i.e., sepals that look like petals. They are leaves (sepals of the calyx) that have been transformed into petals over the course of evolution. They are white on their inner side, like a petal, but their outer side is green like a leaf (see picture in the commentary).

372 He is referring to the green hellebore (*Helleborus viridis* L.), notable for its green flowers.

radice fructus exoritur; omnis enim flos intra caulis corticem continetur. <76> De flore igitur, et de floris tegmine haec in vniuersum dicta sint; de fructu autem, qui sub flore continetur, similiter dicamus.

CAP. VIII.

<77> FRVCTVM vocamus, quod ex semine et semen continentibus corporibus constat: quamuis proprie secundum nominis appellationem ea pars significetur, qua fruimur in cibis expetentes: expetimus autem inter cibos aliquando nuda ipsa semina, vt Pini, Nucis, Castaneae, et omnium frugum et leguminum. Aliquando carnem seminibus circumpositam, quam proprie Pericarpium vocant, vt Mali, Piri, Melopeponis. <78> Cum igitur de semine superius dictum sit, relinquitur, vt de circumpositis corporibus dicamus: hinc sumpto initio. Seminibus omnibus inest humor quidam foecundus, quo euanescente, aut per aetatem, aut ab externa iniuria redduntur infoecunda: Huius igitur custodiendi gratia natura omnibus corticem quendam circumduxit, qui perpetue haeret, donec germinare coeperint; Ob id enim plurima asseruari integra per annum possunt, quaedam longe diuturnius, quod animalibus nequaquam contingit paucis exceptis. Traduntur enim piscium Oua maxime perdurare eo argumento, quod lacus exsiccati per aestatem, aduentante imbre pisciculos ferunt. Papilionum quoque oua per annum durant: minime autem omnium auium oua asseruantur. <79> At plantarum quarundam semina, vel in quadragesimum annum foecunda perdurare proditum est: magis autem perdurant, quae osseo putamine muniuntur, quam quae membranaceo, ob eandem rationem: seminibus igitur mollioribus natura corticem duriorem crassio-

76. [Conclusion on flowers]
This is what is generally said on the subject of the flower and what protects it. Now let us move onto the fruit, which is contained beneath the flower.

Chapter 8 [Fruits and seed coats]

77. [Definition of fruit and pericarp]
We call the fruit that which is composed of the seed and the bodies that contain it, even if more narrowly, this word refers to the part that we seek to enjoy as food. However, sometimes we look for the edible part within the seeds themselves which we strip – as in the case of pine kernels, walnuts, chestnuts, legumes, and everything we harvest (*frux*[373]) –, and sometimes in the flesh which surrounds the seeds, which we specifically call the pericarp – for example that of apples, pears, or melons (*Melopepon*).

78. [Protection and durability of seeds]
As we have dealt with the seed above[374], let us use it as a starting point and then put it to one side, in order to speak about the parts of the fruit that surround it. In all seeds is located a sort of fertile liquid; when it evaporates, whether from age or from an external wound, [the seeds] become sterile. Consequently, in order to protect it, nature has developed around all [seeds] a kind of bark which is retained up until the point of germination. Thanks to this, most [seeds] are preserved intact throughout the whole year, and some even much longer, which is not in any way the case in animals, with some exceptions. Indeed, fish eggs are known to survive for a long time because even if the water where they live dries up in the summer, the return of the rains brings back the fry (*pisciculus*). Butterfly eggs also survive all year long, and yet they are the least watched over of all bird eggs (*avis*).[375]

79. [Texture of piths and barks]
But it is a question of investigating the seeds of plants, that is [those which] remain fertile, even after forty years.[376] In fact, those endowed with a shell (*putamen*) which is of a bony nature rather than membranous last longer for this very reason: nature has given a harder and thicker bark to seeds that are more tender, like walnuts (*Nucle-*

373 *Frux* refers to anything that is harvested in some way, including vegetables and grains.
374 Cesalpino has dealt with the seed in all chapters, but particularly in chapter 6.
375 Does this imply that Cesalpino considers butterflies to be birds? Perhaps the author of *De plantis* uses the term in a wider context to incorporate all flying insects? Varro, *De re rustica* III, 16 (ed. and Fr. transl. Nisard 1864: 148–154) uses the word *avis* at least for bees. He also uses the synonymous *volucris* for bees. However, in more recent editions such as Heurgon and Guiraud 1978–1997: 3.37–38, the word *avis* has been deleted and only the word *volucris* remains to designate bees.
376 This seems to be a way of saying “those that last a very long time”, in contrast with fish or butterfly eggs which last a year at the most.

remque tribuit, vt Nucleis, Pineis, Palmeis, Persicis, et tandem omnibus, quae osseo cortice teguntur: omnium enim medullae molles et pingues sunt, vt mirum videatur, quo pacto in seminis germinatione os adeo durum dehiscat in duas partes dissectum, cum vix ictu mallei rumpatur. Quibus autem duriora et sicciora insunt semina, iisdem cortex tenuior, et mollior datus est, scilicet cartilagineus, aut membranaceus, vt Castaneae, Glandi, leguminibus, frumentisque. <80> Quemadmodum autem in ouo sub cortice testaceo membrana quaedam mollis continetur, mollem foetus materiam tangens, sic seminibus omnibus duplex cortex datus est, internus quidem tenuior, ac mollior, membranaceus medullam vndique tangens, qui tamen in quibusdam facile abscendat,
[17] vt Pineis, in quibusdam pertinacius haereat, vt in Tritico: externus | vero, durior et crassior, vt plurimum osseus, vel cartilagineus: qui facile seiungatur ab interna membrana in plurimis, in quibusdam tamen propter tenuitatem seiungi difficilius possunt, vt in leguminibus et frumentis: intra membranam internam vnam vnicum perpetuo semen continetur: sub cortice tamen externo casu aliquando bina continentur seiuncta per propriam membranam, vt in Amygdala: <81> Nectitur semen cum cortice, qua parte cor est. Cum enim in ea parte sit radicis principium, per eandem quoque veluti per vmbilicum ex parente trahit alimentum: patet autem alimenti ductus ad eam partem tendens; perfecto enim iam semine, a caeteris quidem partibus seiungitur; haeret autem solum, qua cor est pendens aliquando medio quodam peciolo in corde seminis

us),[377] pine nuts (*Pineus*), palm nuts (*Palmeus*), peach kernels (*Persicum*) and all those seeds which are protected by a bony bark. Furthermore, the piths of all seeds are soft and oily, which seems surprising; consequently, in the development of the seed, let us divide the hard stone (*os*) in the way in which it is split into two parts when we break it with a sharp hammer blow.[378] But fruits which contain harder and drier seeds are endowed with a thinner and softer bark, in other words cartilaginous[379] or even membranous, like the chestnut, acorn (*Glans*), legumes and grains.

80. [The two barks]

Just as there is under the protective shell (*cortex*) of the egg a certain soft membrane that is in contact with the even softer substance of the fetus, so all seeds are endowed with a double bark. Firstly, an internal [bark], thinner, softer, membranous, and fully in contact with the pith, which in some species (such as the pine nut) easily disappears, but in others (like wheat) remains in place for a long time. Secondly, there is an outer [bark] | which is harder and thicker, most often bony or cartilaginous, which in most [17] cases separates easily from the internal membrane, although in some is more difficult to separate because it is so fine, as in legumes and cereals. Within one internal membrane we always find one single seed. However, beneath the outer bark, there can be double disconnected [seeds], contained within their own membranes, as in the case of the almond.

81. [Position of the link between the seed and the bark]

The seed is attached to its bark through the part where the heart is situated. Indeed, since the principle of the root is found in this part [the sprout, where the heart is], the food is also drawn through it, just like it is through the umbilicus from the parent, and it is obvious that the food passage[380] extends to this part [that of the heart]. Moreover, from the point of maturation, the seed is effectively separated from the other parts, and is held only by [the sprout] where the heart can be found, somehow attached to

377 Up until this point, Cesalpino has used the word *Nux* for walnut, but here he has shifted to its synonym *Nucleus*.

378 Cesalpino seems to be proposing to establish a (theoretical) division in the development of the seed based on the (physical) division that we can see in a kernel that has been cracked by a hammer. Nevertheless, this passage remains ambiguous. It probably indicates that one can theoretically separate the inner part of the stone (the soft, fatty kernel) from the outer part (the shell) insofar as these two parts come apart when a stone is broken and opened.

379 Cesalpino differentiates three levels of rigidity: bony, cartilaginous, and membranous (from the most rigid to the softest). These terms bring plants closer in line with animal tissues.

380 In current botanical terminology, the feeding tube that connects the seed to the internal wall of the fruit (the ovary) and that is sometimes compared to the umbilical cord is known as the funiculus.

desinente, vt in seminibus Mali et Piri est manifestum. <82> Verum non similiter externus cortex in omnibus secundum eamdem partem cum planta nectitur: Nam quaedam eo modo sedent, vt seminis cor exterius vergat, vt in Castanea, et in plerisque arborum fructibus contingit. In herbaceis vero pluribus idem est nexus seminis et externi corticis, vt in leguminibus et frumentis; cum autem seminis substantia ex interna caulis medulla ortum ducat, necessarium fuit ex intimis germinum partibus neruulos egredi ad fructum vsque, per quos materia seminis ferretur: Hinc fit, vt non fructificent plantae ex ramis crassioribus, paucis quibusdam exceptis, vt Siliqua et Ficus Aegyptia: sed ex surculis, vt plurimum annotinis, vt Amygdala, aut nouellis germinibus, vt Vitis. Nam seminis materia ex his magis erumpere potest, quam ex crassiori ligno: iis igitur neruulis alimentum ferentibus appensa sunt omnia semina cum propriis corticibus: vt neruulorum substantia in corticem transeat, tenuius autem alimentum in semen: Est autem neruulorum nexus cum cortice duriori aliquando in ea parte, cui subiacet seminis cor: aliquando in altera: in his autem necesse est intra duriorem corticem neruulum ad seminis cor ascendere, vt manifestum in Persicis est et in Amygdalis, et omnibus, quorum cor spectat extra. <83> Corticis figura in plerisque pro figura seminis contenti habetur: Nam hoc vel rotundum, vel oblongum, vel depressum, vel recuruum, vel angulosum est. Aliquando tamen aliter se habet, vt in semine Rusci: medulla enim tenuis est, et oblonga vix conspicua ob exilitatem, cortex vero

the central petiole,[381] at the far end of the heart of the seed, as we can see in the seeds of apples and pears.

82. [Configuration of the link]
In fact, the outer bark is not attached in the same way or to the same part in all [plants]. Certain [outer barks] are fixed in such a way that the heart of the seed points towards the outside, like in chestnuts and most tree-growing fruit.[382] But in many herbaceous plants, the heart is a junction point between the seed and the outer bark, as in legumes and cereals. And since the origin of the seed's substance is the internal pith of the stem,[383] the nervures which transport the seed's matter must run from the inner parts to the shoots with the fruit. This takes place so that plants do not fruit from branches that are too thick – with the exception of some like the carob (*Siliqua*) and the Egyptian fig (*Ficus Aegyptia*)[384] – but rather from scions; in most cases the fruit is borne on year-old scions, as in the case of the almond tree, otherwise from new shoots, like the grapevine. Indeed, from there, the seed matter can spread further than from a thicker wood; all seeds, with their respective barks, are therefore attached to these nervures which carry their food, so that the substance passes through the nervures to the bark, and the more refined food passes into the seed. Furthermore, the point where the nervures meet the harder bark is sometimes in the part where the heart of the seed is located, sometimes in another part. In these [parts] the nervure must rise to the heart of the seed by crossing the harder bark, like we see in peaches, almonds, and all those where the heart protrudes beyond the coating.

83. [General appearance of the seed bark]
In most [seeds], the shape of the bark depends on the shape of the seed it contains. Indeed, it is round, oblong, flattened, curved or angular. Sometimes, however, it depends on something else, like in the seed of the butcher's broom: its pith is delicate and stretched out (*oblonga*), hardly visible because of how fine it is, whereas its bark is rounded, thick and bony, to the point where it seems to be made of bone.[385] In other

381 *Medio peciolo*, translated as "central petiole", probably refers to the core of the apple and pear where the seeds are attached, and which extends to the axis of the petiole or the peduncle of the fruit (the "stalk" of the fruit).

382 The heart of the seed is in fact the basis of the (future) sprout; it protrudes from, for example, the chestnut, forming the point from which the shoot will sprout.

383 Cesalpino has in effect demonstrated in chapter 3: §37, that seeds are fed by the pith and not through the bark.

384 The sycamore fig (*Ficus sycomorus* L.), to which he is probably referring here, is mainly grown in Egypt.

385 Here, Cesalpino plays on the word *osseus* (bony/rigid) and *ex osse constare* (composed of bone). The case of butcher's broom seems simply to be an exception to the rule that he has set out, and no explanation is given for the reason for this exception.

rotundus et crassus osseus vt totus ex osse constare videatur. In aliis appendices corticis datae sunt, vnde variatio figurae, vt in Tribulo. <84> Color cuiuscumque generis reperitur perfectis seminibus viridi excepto: sicca enim est valde corticis substantia, herbaceus vero color sine humore non fit in plantis. Nigra et omnino atra multa sunt: corticis enim substantia excrementitia est, non pura, vt contenti seminis. Idcirco semina alba tantum sunt, vt interna plantarum non tacta ab aere ambiente: sic enim et
[18] folia alba redduntur, si obruantur, vt Intubi, Porri, Cardui, | et flores, quamdiu inclusi sunt suis calycibus, calidiores sunt, explicati, vt aere circumfuso fruantur, colores proprios assumunt. Seminum autem cortices ob impuritatem alimenti nigredinem assumere potuerunt, et caeteros colores licet non adeo sinceros, vt in floribus; Vergunt enim aliqua ex parte ad eosdem colores cum floribus, cum eorum materia simul erumpat; recedunt autem: quia purior materia secernitur in semina, tenuior in flores, crassior in cortices; pro diuersitate autem externi corticis denominantur ipsa semina alba, vel nigra, russa, punicea, aut alia huiusmodi; quoniam hic perpetue haeret semini. <85> Cortices igitur, qui semen custodiunt ad germinationem vsque, hi sunt in omnibus.

[seeds], the bark is endowed with appendages, to the extent that its shape is altered, as in the goat's head (*Tribulus*).[386]

84. [Color of the bark]

The color [of the seed bark] of the species[387] reveals itself once the seeds reach maturity, except for green: greenness does not exist in plants without sap (*humor*); for the substance of the bark is completely dry. Many of them are black and completely dark, because the substance of the bark is excremental, and not pure like [that of] the seed which it contains. It is for this reason that seeds are so white: as they are inside the plants, they are not touched by the ambient air. And it is for the same reason that leaves whiten if they are buried, like Belgian endive (*Intubus*), leeks and cardoons (*Carduus*),[388] | as do the flowers as long as they are enclosed in their calyxes; at the [18]
point at which they benefit from the circulation of air, they warm up and unfurl, and then acquire their own colors. The seed barks perhaps owe their blackness to the impurity of their food, and in view of this they are not able [to receive] their own colors, like flowers do. Indeed, to a certain extent they tend to have the same colors as flowers, since the development of their matter happens at the same time.[389] But they [the colors] fade because a purer matter is distributed to the seeds, another finer one to the flowers, and a thicker one to the bark. Depending on the variety of the outer bark, the seeds themselves are referred to as white, black, red (*ruffus*), purple (*puniceus*)[390] or other similar colors, since that [this color] remains permanently in the seed.

85. [Conclusion on seeds bark]

Here then are all aspects of the barks which protect the seed until its germination.

---

386 See illustration in the commentary.

387 The word used is *genus*, but probably not in the sense defined in chapter 12 and used from there on (to distinguish between trees, shrubs, undershrubs, and herbs) but rather in the sense of a species, as was already the case in chapter 2. This interchangeable use between genus and species exists in Aristotle, but less so in Aristotelian authors who came after Porphyry, who respect more systematically the difference between genus and species.

388 Out of context, the term refers to either the cultivated artichoke, or a sort of edible thistle, in other words the cardoon, or a thistle of that kind. Considering the crop and food context here, *carduus* is most likely referring to an edible species.

389 The bark develops at the same time as the matter from which the flowers will be made, so we can assume that their natural colors are the same, even though impurities darken the bark.

390 *Rufus* or *russus* (that Cesalpino combines here in *ruffus*) means red or reddish-brown, whereas *puniceus* means red or crimson/pink/mauve, hence our choice of translation.

CAP. IX.

<86> QVAMVIS autem inuolucra, quae diximus, sufficere videantur ad perfectorum seminum tutelam; non sufficiunt tamen ad eorumdem perfectionem: Nam antequam efficere possint talia, qualia ipsa sunt, egent concoctione quadam perficiente; quoniam autem concoctio a calore interno fit, humorem seminum incrassando: Initio enim semina omnia humida sunt, et ferme aquea: pefecta autem constant et sicciora sunt reddita, oportuit calorem innatum foueri in seminibus, ne exterius transpirando, ipsa euanida redderentur. Huius igitur gratia natura semina obduxit alio corpore, vt plurimum carnoso et humecto: quemadmodum Medici tuberculis concoquendis cathaplasma aliquod superponere solent, vt humiditas cum calore in parte conseruetur. <87> Appellatur autem in fructibus haec pars pericarpium; quia pulpa est exterius obducta, quae in maturitate cibis expectitur in plerisque vt in Vua, Cerasiis, Piris: Dum igitur maturatur pericarpium a calore interno, adiuuante tamen extrinsecus sole, semina interius recondita perficiuntur: simul enim ab eodem calore pericarpij humor post acerbitatem, aut aciditatem per maturationem acquirit dulcedinem, et semina ad perfectionem accedunt, vt sibi similia gignere possint. <88> Non est autem eadem seminis materia et pericarpij; illa enim ex profundis partibus erumpit, vt superius comprobauimus: haec autem ex exterioribus magis accedit, scilicet ex cortice. Rationi enim

Chapter 9 [The pericarp]

86. [Importance of coction]

Although the husks, which we have dealt with,[391] seem to suffice to protect seeds that have reached maturity, they are not, however, sufficient for maturation. Indeed, before they [the husks] can get them [the seeds] to what they will be, they [the seeds][392] lack the coction (*concoctio*)[393] which drives them to maturity: when this coction is produced by inner heat, it is in a position to thicken the seeds' liquid. Indeed, all seeds start off being damp, almost aqueous; once they have reached maturity, they stabilize and become drier; [before that] the innate heat has had to be preserved in the seeds, so that it does not disperse (*transpirando*) to the outside and weaken them. That is why nature covers the seeds with another body (fleshy and moist in most cases), in the same way that physicians are accustomed to applying poultices to warm (*concoquo*) tumors, so that the moisture is preserved with the heat in this part.[394]

87. [Protective function of the pericarp]

In fruits, we call this part the pericarp, because it contains the pulp, which in many fruits is the part that is sought after to eat when ripe, like in grapes (*Uva*),[395] cherries (*Cerasum*) and pears. From then on, until the pericarp ripens by means of internal heat (nevertheless as well as with the external assistance of the sun), the seeds mature by staying hidden inside. Thanks to this same heat, the pericarp's liquid simultaneously acquires a sweetness, after the bitterness or the acidity it had during maturation, and the seeds reach their maturity, in order to be able to produce more [seeds] like themselves.

88. [Origin of the pericarp]

The matter of the seed is not the same as that of the pericarp: the former comes from deeper parts, as we have already shown;[396] the latter more from the outside, in other

---

391 See chapter 7 (§64–75) for what Cesalpino described as husks (*involucrum*), but which perhaps also includes seed barks that he has just dealt with in chapter 8.

392 Cesalpino has already alluded to this process in chapter 6 (§61): the dry seeds, buried in the soil, absorb the nutritive liquid from it, which awakens their inner heat, and this prepares and "cooks" (*concoquo*) the liquid.

393 In Latin and in English, the term largely signifies a transformation by heat, but in the medical vocabulary of the time, it deals more precisely with the change undergone by the bodily humors.

394 In other words, the body in question (the pericarp) serves as a "poultice" to the seed, to prevent the heat from escaping too quickly, taking with it the moisture, and thus drying out the seed.

395 In Book III, chapter 18, the term clearly indicates the grape, from which we make wine. Greene (1983: 2.568) explains that Latin botanical texts distinguish the names of plants from the names of their fruits. Sometimes the nuance is thin, for example between *pirus* for pear tree and *pyrum* for the fruit, or *cerasus/cerasum* (cherry tree/cherry); sometimes it is more important, as in the distinction between vine (*vitis*) and grape (*uva*) which he cites as an example.

396 See chapter 3 (§37) and chapter 8 (§81–82).

consonum est, vt quemadmodum pars neruea cum medulla cortice tegitur in germine, sic in fructu, quod ex internis oritur tegatur ab aliquo corpore, quod ab externis ortum ducat, et vt ibi tria sunt scilicet, Medulla, lignum, et cortex: sic in fructu, semen, cortex, et pericarpium, aut aliud quid, quod pericarpij vicem gerat. Indicio autem est, quod pericarpium materiam ex cortice sumat; nam inoculatio fructum parit pro natura corticis inoculati. Semen autem pro natura ligni, vt superius notauimus. Nec obstat, [19] quod flos externus | sit, qui tamen ex internis proueniat: quando enim flos circundat, pericarpium adhuc non est factum, sed vestigium tantum adest tum seminis tum reliquarum partium: postquam vero flos deciderit, iam incrementum omnia assumunt. <89> Praeterea nil mirum videatur, si floris materia ob tenuitatem erumpere extra pericarpium possit, cuius tamen materia non ex internis proueniat. Pericarpium omne ante maturitatem coloris herbacei est, vt germinum cortex: iis exceptis, quae altero cortice integuntur, vt Punica: Nam quae aeri non sunt exposita, alba sunt: postquam vero maturata sunt, omnia proprios colores acquirunt, sed exterius magis quam interius, et maxime qua solem spectant. Nam vel flaua redduntur, vel punicea, vel purpurea, vel nigricantia, vel alba quaedam etiam caerulea, vt Lauri syluestris baccae, dulciora quoque redduntur et odoratiora per maturationem. <90> Constant quaedam magis humore, vt Vua: quaedam carne vt Ficus: Quaedam duriori pulpa vt Poma: inest quibusdam succus aquosus, qui facile putrescit, vt Cerasiis, Persicis, et tandem fructibus aestiuis, quos Orarios vocant: Quibusdam melleus et hi exsiccati diutius conseruantur, vt Ficus, Vuae. Quibusdam pinguis, ex quibus oleum extrahitur, vt ex oliuis, quibusdam exiguum inest pericarpium, vt Laurinis baccis, et tandem syluestribus omnibus: Contra multo abundant pericarpio, quae aquosa sunt, vt Cucumeres, et tandem fructus domestici: Aliis pericarpium singulis seminibus proprium datum est, vt Puni-

---

13 *integuntur*] *in teguntur* impr.; corrigimus. **21** *vt Ficus Vuae* impr.; corrigimus.

words from the bark. This conforms to reason, because in the same way that the fibrous part in the shoot is protected by the bark, what comes from the inside within the fruit also needs to be protected by another element that draws its origin from the outside. From that we can see that there are three things, namely the pith, the wood, and the bark; likewise in the fruit, the seed, [its] bark and the pericarp, or something else that serves as a pericarp. The proof is that the pericarp receives its matter from the bark [of the tree]; indeed, shield budding will produce the fruit from the bark that is grafted. The seed [receives its matter] from the nature of the wood, as we have remarked above.[397] The fact that the flower is on the outside [of the seed] is no objection. | Indeed, it still comes from the inside, because when the flower surrounds [19]
it, the pericarp is not yet finished, but a trace – both of the seed as well as of the remaining parts – is present; and as soon as the flower falls, all [parts] start to grow.[398]

89. [Color of the pericarp]

Furthermore, we should not be surprised if the matter of the flower, because of its delicacy, can emerge on the outside of the pericarp whose matter, in contrast, does not come from the inside. Every pericarp has a grassy color before its ripening (like the bark of the shoot), with the exception of those which are covered in another bark, like the pomegranate. Indeed, these pericarps are not exposed to the air; they are white. However, after their ripening, all acquire their own colors, but on the outside more than on the inside, and especially those that see the sun. Indeed, they become either orangey, red, purple, black, white, or even blue-tinged, like the berries of the laurustinus (*Laurus sylvestris*) and become sweeter and more fragrant once they are ripe.

90. [Composition of the pericarp]

Some [fruits] are made up of more liquid, like the grape; some of more flesh, like the fig; and some of a harder pulp, like the apple (*Pomum*).[399] Some contain a watery juice, which rots easily, like cherries, peaches and all those summer fruits that are referred to as "coastal" (*Orarios*). Some [contain a juice that is] honey-like; once they have been dried for a whole day, they keep well, like figs and grapes. Some [contain a juice that is] oily, from which oil is indeed extracted, such as olives. Some have a very fine pericarp, such as laurustinus berries and also all wild (*sylvester*) [berries]. On the other hand, watery [fruits] are filled with a plentiful pericarp, like cucumbers (*Cucumis*) and cultivated fruits. In others, each individual seed is endowed with its own pericarp, like the pomegranate. And in others, several [seeds are endowed with a pericarp] in common,

---

397 See chapter 3 (§37) and chapter 8 (§81–82).

398 Literally they "receive their growth", which probably means from the matters from which the fallen flower no longer draws, which allows them to grow better.

399 Up until this point, Cesalpino has used the word *Malum* to refer to the apple tree and its fruit, but here he uses the synonym *Pomum*.

cae. Aliis commune pluribus, vt Melopeponi, semina in quibusdam seiuncta a pericarpio membranis intercedentibus, ut in Pomis, et tandem iis, quorum semina cortice non osseo, sed membranoso constant; ut enim in foetu humor excrementitius inter membranas continetur, vt foetus in sicco sit non in humido; sic semina maxime in sicco esse volunt; in quibusdam humor pericarpij vtriculis continetur, quibus stipantur semina vt in Limonibus, quibus et duplex genus pericarpij datum est unum succosum iuxta semina, alterum carnosum exterius loco uasculi. <91> Syluestrium fructus ad maturationem minus perueniunt ob siccitatem: cum tamen eorum semina non minus perfecta sint, fit enim in domesticis aliquando tanta pericarpij accessio, ut semina reddantur uana, aliquando solum insit seminis uestigium, ut in Punicis dulcibus saepe contingit. Quibus autem pericarpium est natura siccius, non maturatur, vt in Amigdala, et Nuce, a quibus abscedit, cum semen perfectum est, maturatio enim in humore fit, post maturationem autem pericarpium omne exsiccatur in membranosam substantiam, et tabescit, ut patet in Cucurbita, et Mandragorae fructu seminibus integris permanentibus; si uero putrescat prius, periculum est ne semina contagio afficiantur, et pereant, putrescunt autem imbribus superfluis, uel si laedantur ictu, ideo in arbore melius exsiccan-
[20] tur a sole, si caetera tempestiue succedant. Decerpti | autem fructus quidam magis, quidam minus putrescunt; idcirco qui semina colligunt, eximunt a pericarpio, antequam putrescat, et in sole siccant. <92> De pericarpio igitur haec satis.

---

1 *Melopeponi*] *Melopoponi* impr.; corrigimus. **11** *siccius*] *sic siccius* impr.; C. corrigit

like the melon.[400] In some [of these fruits], like apples, and also those whose seeds are made of a bark which is not bony but rather membranous, the pericarp has dividing membranes separating the seeds. Furthermore, just as in the case of the fetus, where the excremental liquid is contained by the membranes while the fetus is kept dry and not in the wet, seeds too want to be as dry as possible. In certain [fruits], the pericarp liquid is contained within little pockets (*utriculus*),[401] where the seeds are tightly packed, like in citrus fruits (*Limus*),[402] which are endowed with a double type (*genus*) of pericarp, one full of juice and surrounding the seeds, the other fleshy and on the outside of where the capsule can be found.[403]

91. [Rotting of seeds and pericarp]

Wild fruits are less successful in ripening in the case of drought, although their seeds are nonetheless mature. In cultivated [fruits], the pericarp sometimes shrivels (*accessio*) to such an extent that the seeds become empty and sometimes only a trace of the seed is left, as happens often to sweet pomegranates (*Punica dulcis*). In these, the pericarp is drier by nature, so that it does not ripen, as in the case of the almond and the walnut from which it detaches when the seed is mature, since maturation happens in the liquid, and after the maturation, the whole pericarp dries up into a membranous substance and spoils, as we can clearly see in marrows and the fruit of the mandrake (*Mandragora*), where the seeds remain intact. But if it [the pericarp] rots earlier, the risk is that the seeds will be contaminated and die; they rot if there is heavy rain or if they are damaged by a blow. That is why, in trees they dry out better, thanks to the
sun,[404] if the sequence of circumstances happens favorably. And once picked, | some [20]
fruits rot quicker than others; that is why those who harvest seeds take them out of the pericarp before it rots and dry them in the sun.

92. [Conclusion on the pericarp]

That is sufficient with regard to the pericarp.

---

400 We are assuming *Melopeponi* rather than *Melopoponi* (the Italian translation) used in the text.

401 These are most likely the juicy segments of citrus fruits that contain their pips.

402 Cesalpino uses the Italian term *Limus* here, rather than the more common *Citrus*.

403 *Vasculum* most likely refers in Cesalpino's work to any container of seeds, and not only a dry fruit capsule used in more limited current botanical terminology. In contemporary terminology, the second pericarp is therefore the zest, in other words the epicarp, whereas the first pericarp around the seeds is the mesocarp.

404 They benefit from being dried by the sun, because that prevents rotting through an excess of moisture.

CAP. X.

<93> NON omnibus autem datum est pericarpium; nullis enim Frumentis, neque leguminibus, neque vt summatim dicatur generi herbaceo, paucis quibusdam exceptis, vt Solano, Mandragora, Fragaria: arboribus autem et fruticibus plerisque datum est. Eo autem carent Pinus, Abies, et reliquae coniferae, item Auellana Fraxinus, Glandiferae omnes: quibus autem pericarpium negatum est, iis aliud corpus, quod eius vicem gerat, tributum est, siccius quidem pericarpio, sed non durius seminis cortice; oritur enim, vt pericarpium ex extima parte germinis, et eandem praestat semini vtilitatem. Quoniam vero eorum natura siccior est, siccius quoque corpus seminibus fouendis concessum est: <94> Nulla autem sunt, quae aut pericarpio, aut proportionali tegmento careant vsque ad seminis perfectionem: nisi forte pauca, quibus satisfaciat corticis crassities et durities, vt Tribulus, Lithospermon; in his enim videntur nuda omnino semina in caule ferri: extra corticem enim lapideum nullum est aliud inuolucrum; sunt tamen vmbilici quidam florem tegentes, in quibus multo tempore clauduntur semina, perfecto autem semine, resupinantur, et nudum id ostendunt, <95> in foeniculo et caeteris ferulaceis, teguntur semina quadam cute caulis striata: nam in quibusdam eius generis manifestus est nucleus internus durus, vt in candido Rosmarino; creduntur tamen in omnibus huiusmodi nuda esse semina, quia externa cutis non abscedit a cortice ob exilitatem, et similitudinem substantiae: extima enim caulis ferulacei non

---

4 *fruticibus*] *fructibus* impr.; C. corrigit.

Chapter 10 [The parts dedicated to the protection of the fruit and fructification]

93. [The dry body]

Not all [plants] are endowed with a pericarp: cereals are not, nor legumes, nor – generally – herbaceous plants, except some like the nightshade (*Solanum*),[405] mandrake or strawberry (*Fragaria*). On the other hand, trees and many shrubs are endowed with one. But the pine, fir, and other conifers (*conifer*)[406] do not possess one; nor does the manna ash (*Avellana Fraxinus*)[407] or any of the glandiferous trees (*Glandifer*). Plants that have been deprived of a pericarp have received another body that serves the same purpose, actually drier than a pericarp, but less hard than the bark of the seed. It comes, as the pericarp does, from the outermost part of the shoot, and has the same purpose for the seed. And since the nature of these [seeds] is drier, a drier body has also been granted to seeds which need to be warmed up.[408]

94. [The bark of the seed]

Until maturity, no seed lacks either a pericarp or any equivalent protection, except a few whose hard and thick bark is enough, like the goat's head and the common gromwell (*Lithospermon*). Indeed, in these, the seeds seem to be held completely bare on the stem: no other stony (*lapideus*) husk is present apart from the bark. In contrast, kinds of lids (*umbilicus*) in which the seeds are enclosed for a long time protect the flower, and once the seed reaches maturity, they retract and unveil it.[409]

95. [Bare seeds]

In fennel and other Ferulaceous plants, the seeds are protected by a sort of striated skin [which comes] from the stem. Indeed, in certain [seeds] of this genus, a hard inner kernel is visible (*manifestus*), like in pale rosemary (*candidus Rosmarinus*). However, insofar as the external skin does not detach from the bark because it is tightly attached and made of a similar substance,[410] the seeds are, for this reason, considered bare in all [plants]. Indeed, the outermost [parts] of the stem in Ferulaceous plants are not

---

405 *Solanum* is a genus that is used to cover several species categorized under the common name of nightshade.

406 The lack of capital letter could be a typographic misprint, or it could indicate that Cesalpino does not consider the term to designate a genus in itself.

407 *Avellana Fraxinus* is not found like this in any other place in *De plantis*. *Nux avellana* has been previously used to refer to the common hazel, so we can therefore surmise that he is referring here to the fruit of what he calls *Fraxinus*, that of the manna ash (*Fraxinus ornus* L.).

408 At the beginning of chapter 9 (§86), Cesalpino attributes to the husks and to the pericarp the function of preventing the escape of heat during the seed's maturation.

409 What he means by this *umbilicus* is not clear. Perhaps it refers, in the common gromwell, to the protection of the sepals which cover the immature flower bud, which is like a lid.

410 The skin of the bark in question is either the lemma surrounding the grain, which is a part of the chaff, or the seed's integument, the protective husk which constitutes the bran (see illustrations in the commentary).

multum distant ab intimis, cum cortex abierit in folia, Hordei quoque granum, et Zeae nudum videatur: quia vtriculis quibusdam clauditur florem tegentibus, qui difficilius abscedant. <96> Quibusdam igitur sufficit tegmen flori superpositum, vt in Marrubio, Buglosso, Tritico, quo deficiente, abutitur quandoque natura ipsis foliis, vt in Milio Indico, nam paniculae integrae foliis inuoluuntur, cum singulis seminibus neque flores, neque florum tegmina data sint. similiter autem in Tritico, et caeteris spicam ferentibus, aliquo tempore spica foliis est inuoluta, sed postea erumpit. <97> Aliis autem natura praeter floris tegmen proprium inuolucrum seminibus tradidit; idque aut singulis singula, vt in Auellana, Glande, Nuce Vnguentaria, aut vnum commune pluribus seminibus, quasi vasculum multis modis formatum; nam vel oblongum corniculi modo, quae siliqua appellatur, vt in leguminibus, vel rotundum capitis modo, vt in Lichnide,
[21] vel alia qualibet fi|gura, idque vel vnica communi cauitate, vt in Rosa, vel in plures loculos partita, vt in Papauere: aliisque variis modis. <98> Quae distinctio etiam in

very far from the innermost, since the bark extends into the leaves. In the same way, the grain of wall barley (*Hordei granum*), like [that of] spelt (*Zea*), seems bare, because it is enclosed in small pockets which cover the flower, and which come away with difficulty.[411]

96. [Foliar protection and ears]
In some, therefore, an additional protection (*tegmen superpositum*)[412] for the flower is enough, like in the horehound, bugloss (*Buglossum*)[413] and wheat; whenever it is missing, nature makes maximum use of the leaves themselves, like in sorghum: the whole panicles are enveloped by the leaves,[414] because the individual seeds are not endowed with flowers nor with floral protection. In a similar way, in wheat and others [species] which bear an ear, for a certain length of time the ear is enveloped by the leaves, but then subsequently emerges out of them.

97. [Seed husks]
In others, nature gives seeds a husk dedicated to protection in addition to the flower; and this is [by giving] either individual [husks] to individual [seeds], which is the case for the hazel, the acorn and ben nut (*Nux Unguentaria*),[415] or one single [husk] for several seeds together, in other words a capsule (*vasculum*) formed in several ways. Indeed, it can be oblong in the form of a little horn, which is called a silique, like in legumes, or rounded like a head, like in common corncockle (*Lichnis*),[416] or any other shape | either with one common cavity like the rose, or subdivided into several com- [21]
partments like the poppy, and in various other ways in others.

411 These little pockets are probably the glumes and lemma which surround the floret of the ear and then the grain until it matures. In this paragraph, Cesalpino mentions bare grains. In current terminology, we say that the grain is bare when its pericarp, made of cover glumes or husks, detaches after threshing, which is not the case for hulled grains (which require a separate hulling operation). Cesalpino's terminology seems to be different, since here he uses "bare" for grains whose husks are not easily removed.

412 This *tegmen superpositum* perhaps refers to involucre bracts in the current botanic sense.

413 Here Cesalpino uses the term *Buglossum* and not *Anchusa* as he did earlier, which can only refer to the species that he calls *Buglosa hispana* which turns out to be green alkanet (*Pentaglottis sempervirens* (L.) Tausch ex L. H. Bailey).

414 Although he seems to be mentioning sorghum (see note in chapter 7, §72), his description here does not correspond very well to that species. Indeed, unlike an ear or corn, for example, sorghum panicles are not enveloped by leaves.

415 Cesalpino writes that "*Glans unguentaria*" is commonly called "Ben" (Book 2, ch. 45: 79). The ben nut, particularly used in perfumery, according to the identification proposed by the *Nouveau dictionnaire d'histoire naturelle* (1818: 22.150), is the fruit from the *Moringa oleifera* Lam. According to De Vos 2010, *Nux Unguentaria* is the nutmeg (*Myristica fragrans* Houtt.).

416 Most probably common corncockle (*Agrostemma gitagho*).

pericarpiis reperitur: sunt autem quaedam, quae et vasculum habeant et floris tegmen vt Polemonia, Lichnis. Quaedam praeterdicta lanosam quandam substantiam habent, qua semina farciuntur et fouentur, vt Asclepias, Rhododaphne, Populus, Cicoracea, quibus Pappi seminibus insident, veluti plumae quaedam: proportione autem respondent haec iis, quae in vasculo aliud genus pericarpij habent, vt Aurantia, Punica; differunt autem: quoniam pericarpium externum magis videtur, scilicet a tunica germinis ortum: Pappi autem, e cortice seminis efflorescere videntur, vel in summo tantum, qua etiam flosculi in Cicoraceis exoriuntur: vel vndique vt in Rosarum seminibus; est enim valde tenuis et sicca lanuginis substantia, et vt plurimum euolat cum semine relicto vasculo: feruntur enim in auram huiusmodi semina, et volitant ob lanuginem ad iunctam. <99> sedent autem in vasculo semina suis neruiculis adnexa; vel in ambitu, vt in Ficu; vel in medio, vt in Lichnide; vel ab uno latere, ut in leguminum siliquis; uel in duobus lateribus, ut in Chelidonio; uel in inferiori parte, ut in iis, quorum uasculum est floris tegmen, uel in superiori, ut semina deorsum inclinentur superius appensa, quod in plerisque reperitur. <100> Sunt, quorum uascula resupinantur, ideo semina exterius uergunt, ut in Chamaemelo, idem uidetur contingere in fructu Mori, et tandem omnibus, quibus semina exterius aceruata sunt, ut Rubo, Fragaria, Ranunculo: Mori enim fructus est ueluti Ficus inuersa, ut pulpa in medio sit, in ambitu semina,

98. [Capsules and down]

We can also observe this difference in pericarps. Certain [species] have a capsule and a floral protection,[417] like the bladder campion and the common corncockle. Some, as we have alluded to, have a certain downy substance, where the seeds gather and stay warm, like the white swallow-wort (*Asclepias*), the oleander (*Rhododaphne*), the poplar (*Populus*)[418] and the Chicoraceous plants (*Cicoraceum*); in these seeds we find pappi (*Pappus*), which are like feathers. These [seeds] correspond analogically to those encapsulated ones[419] which have a different kind of pericarp, like the orange (*Aurantia*) and the pomegranate, but differ in that their external pericarp seems thicker, insofar as it draws its origin from the tunic of the shoot.[420] As for the pappi, they seem to open up from the seed bark – either in clusters similar to the little flowers[421] in the chicoraceous plants, or scattered like the seeds of roses.[422] Indeed, their substance [of pappi] is downy, rather fine, and dry. In most cases, it blows away with the seed, leaving behind the capsule. The seeds are carried by the wind in this way and fly thanks to the down which is attached to them.

99. [Position of the seeds within the capsule]

The seeds, attached to their fibrils (*nerviculum*) are housed in the capsule, either all around it like in the fig,[423] or in the center like the common corncockle, or on one side like in legume siliques, or on both sides like the celandine (*Chelidonium*), or in the lower part like those [plants] where the flower capsule gives protection, or in the upper [part] like in those [plants] where the seeds attached to the top hang downwards, which one observes in most cases.

100. [Anatomy of the capsule]

Some capsules are inverted, so that their seeds turn outwards, as in chamomile – the same thing seems to happen in the black mulberry (*Morus*) – and also in all those plants whose seeds accumulate on the outside, like the raspberry (*Rubum*), the strawberry, and the buttercup. Indeed, the fruit of the mulberry tree is like an inverted

417 What follows (chapter 10) helps to clarify that the term "floral protection" refers to the protection formed by the sepals of the flower's calyx which remains attached to the fruit.

418 *De plantis* describes two poplars: the white poplar (*Populus alba* L.) and the black poplar (*Populus nigra* L.).

419 As in the previous chapter, the term *vasculum* incorporates the capsule and the segment of a citrus fruit.

420 He is referring, most probably, to the outer bark of the stem, which he covered in chapters 7 and 8.

421 These "little flowers" (*flosculus*) are referred to here in connection with those of the chicoraceous genus in which Cesalpino also places the violet (chapter 12: §132) which is a simple rather than a composite flower; this therefore suggests a non-technical use of the term (this follows on from the footnote at the first occurrence of the term: see chapter 7: §70).

422 The seeds of dog roses (wild roses) are in fact downy (see illustration in the commentary).

423 This is a strange example to give, since in the fig the seeds are distributed throughout the pericarp.

similiter se habent inter se Rosae fructus et Rubi. Vascula vt plurimum duplici corpore constant, externa quidem cute molliori, interna uero duriori opposito modo, quam in seminibus: quod recta ratione fit; cum enim haec a cortice caulis oriantur, hic autem duplex sit in plerisque, et externus quidem mollior, internus uero durior, qui in arboribus liber appellatur. <101> merito eadem ratio respondet in fructu: patet autem id in siliquis omnibus; tenera enim est pars externa, adeo vt in quibusdam mandatur cocta, vt in Pisis, et Phaseolis, in quibusdam etiam cruda, cum sponte maturetur in dulcem substantiam, vt in Siliqua arbore. interna autem pars membranosa est et dura: aliquando cartilaginea, ut cum dehiscit perfectis seminibus, ea longius proiciat cum crepitu, ut Sparto contingit, sed maxime omnium Ricino. Oxys quoque longius expuit semina, et Catanance, ut difficile sit colligere. <102> Cucumis syluestris alia ratione longius una cum seminibus expuit humorem pericarpij: Quia uasculum quasi supra modum tensum abscedente fructus pediculo per ostiolum exprimit, quicquid intus est, ubi maturus est fructus: seiungitur autem pediculus eo tempore minimo tactu. <103> Op-
[22] posito autem modo se habent uascula, quae pericarpio constant; du|rior enim est pars

---

1 *et Rubi uascula, ut* impr.; C. corrigit. **9** *cartilaginea*] *carthilaginea* impr.; corrigimus. | *proiciat*] *proiiciat* impr.; corrigimus.

fig, since its pulp is in the middle and its seeds all around it. The same is true of the fruit of the rose and the raspberry. The seed capsules, in most cases, consist of a double body: the outer skin is softer, and the inside is harder, which is the opposite of seeds. This conforms to reason, since they draw their origin from the bark of the stem, which is twofold in most cases: the outside is softer and the inside, called the bast[424] in trees, is harder.

101. [Siliques and dehiscence]
Fruit follows logically the same matching relationship. This is clear in all siliques. On the one hand they have an outer part that is tender to the extent that some of them are edible when cooked, like peas (*Pisum*) and beans (*Phaselus*), and in other cases [can be eaten] raw, since they ripen naturally (*sponte*) into a soft substance, like in the carob tree. On the other hand, siliques have a membranous and hard interior part, sometimes cartilaginous, so that when they fall, once the seeds have matured, they are thrown far with a snap, as happens to the Spanish broom (*Spartum*),[425] and the castor bean being a prime example. The creeping woodsorrel (*Oxys*)[426] also projects its seeds far, as does the touch-me-not balsam (*Catanance*)[427], to the point where they are difficult to collect.

102. [Dehiscence in the squirting cucumber]
When the fruit of the squirting cucumber (*Cucumis sylvestris*)[428] is ripe, it expels the liquid of its pericarp in a different way at the same time as its seeds. Indeed, when the peduncle of the fruit detaches, its capsule, stretched so to speak beyond capacity, expels all that it can contain through a small orifice, when the fruit is ripe. At that moment, the peduncle breaks at the slightest contact.

103. [Non-dehiscent fruit]
[Some fruits] have capsules which remain attached to the pericarp, unlike the species we have already mentioned. Indeed[429], | their external part is harder, as in the case of [22]

424 As already explained in chapter 3: §32.
425 The name spartum *stricto sensu* refers to *Lygeum spartum* L. (esparto grass). However, Cesalpino (book III, ch. 35) also discuss "Spartium" whose noisy dehiscence better fits the description here. Caruel (1853: 159) identified *Spartium* specimens in Cesalpino's herbarium as *Spartium junceum* (Spanish broom, rush broom, weaver's broom) or as *Genista radiata* (southern greenweed, rayed broom). In addition, these two species grow in Italy, unlike the first one.
426 See book XV, ch. 2 (which cites Pliny, *Naturalis historia*, XXVII, 89 (§112); see also XXI, 69 (§112–113) (ed. and Fr. transl. André et al. 1947–2015: 27.58 and 21.67, respectively; ed. and Engl. transl. Andrews et al. 1938–1962: 7.450 and 6.243–245, respectively).
427 *Impatiens noli-tangere* L. owes its name to its capsule's habit of exploding when it is touched, casting its seeds several meters away.
428 This squirting cucumber or exploding cucumber, *Ecballium elaterium* (L.) A. Rich. as it is named today, is a common species in Europe and especially in the Mediterranean region.
429 "Indeed" because the opposition relates to the hardness or softness of the external skin.

externa, ut in Cucurbita, Punica, similiter se habent, quae solitaria sunt, Persica, Pruna: iis enim cutis durior exterius circumposita est ad tutelam pericarpij mollioris, quibus superdatur floris tegmen, nisi constent pericarpio, <104> uasculum, ut plurimum cartilagineum est totum, ut Lichnidis, Polemoniae. Est autem unus fructus, qui uno tegmine continetur externo, quamuis interna diuisa sint; ut in Peonia, et Aconito, tres siliquae aut plures in eodem exortu sub eodem flore continentur; quae enim sub eodem flore sunt, sunt etiam sub eodem floris tegmine, quod externum est. <105> Plures uero qui nullo communi tegmine teguntur, licet ex eiusdem sedis diuisione oriantur, ut Vuarum acini; singuli enim sub singulis floribus sunt, licet racemus unus sit. <106> Ambigunt quidam, vt partim vnus videantur, partim plures, vt Frumentorum spicae: cum enim vnum sit initio, totius spicae tegmen ex caulis foliis vnus videatur fructus: cum vero in spica singulis seminibus singuli flosculi et membranulae flosculos tegentes dati sint, plures videantur in vna spica fructus. <107> An si tegmina ex foliis consideremus, omnia huiusmodi vnus essent fructus non plures ? Nam omnes racemosi et spicosi fructus in exortu foliis teguntur: et tandem quaecunque in nouello germine fructificant, fructum sub foliis reconditum initio ferunt, sed quoniam folia alterius finis gratia data sunt, flores autem et florum tegmina solius fructus gratia: merito ex his non ex foliis

the marrow and the pomegranate; certain similar [fruits for their part] have individual [capsules] like the peach and plum: their outsides are covered by a harder skin[430] to protect the softer pericarp[431] and are endowed with a floral protection on top of that, unless they are attached to the pericarp.[432] More often than not, the capsule is wholly cartilaginous, like that of the common corncockle or the bladder campion.

104. [One floral protection for several pericarps]
A single fruit is contained by a single external protection, although the internal [protections] are divided, like in the paeony (*Peonia*) and the aconitum (*Aconitum*), [where] three siliques of the same origin, or more, are contained beneath a single flower. Indeed, they are [not only] beneath the same flower, but they are also beneath the same floral protection, the one which is on the outside.

105. [One cluster but several flowers]
On the other hand, many [fruits] do not share any common protection, even if their division comes from the same implantation, like the seeds of grapes. Indeed, individual [seeds] are found beneath individual flowers, without this preventing them from forming a single cluster.

106. [Question of the individuality of the fruit]
Some people are indecisive with regard to the fact that [these fruits] appear both to be one single [fruit] and at the same time several, like ears of grain. Indeed, the totality of the protection of the ear seems to [constitute] a single fruit from the leaves of the stem, in the way that it initially seems to constitute a unique entity. But to the extent that within the ear, the individual seeds are endowed with individual flowers, as well as little membranes (*membranula*) which protect these flowers, several fruits seem [to exist] in one single ear.

107. [Based on the leaves or the flowers?]
If we consider the protection from the leaves, would not all fruits form like this as one, rather than many? Indeed, all bunches and ears protect the fruit in the first place using leaves. And everything that bears fruit in a new shoot first carries the fruit hidden under the leaves, but since the leaves are dedicated to another purpose, whereas the flowers and the floral protections [are dedicated] uniquely to the protection of the fruit, it is right that the singularity or the plurality of the fruits should be examined on the basis of them [the flowers and their protection] rather than of the leaves.

---

430 This refers likely to the epicarp, the skin of the fruit.

431 Probably the mesocarp, the flesh of the fruit.

432 The floral protection that he mentions no doubt refers to the sepals remaining from the flower calyx that are on top of the fruit, and which are more or less adnate depending on the species.

vnitas et multitudo fructuum spectanda est. <108> si qui autem flores sunt singulis seminibus distributi, sed sub vno communi florum tegmine, vt in Acanaceis spectatur, vnus est fructus dicendus non plures: si autem nullum sit commune tegmen, licet flores racematim cohaereant et fructus, vt in Ligustro et Vite, non vnus est dicendus fructus: sed fructuum racemus. Est autem vnus racemus, qui vno principio ex caule finditur in plures fructuum sedes. <109> Non tamen in omnibus id racemum vocant, sed in quibus fructus pericarpio constant, vt in Vite: in frumentorum genere spicam vocant, quae erecta est; Iubam, quae sparsim propendet vt in Milio; Paniculam, quae magis compacta, vt in Panico; in genere ferulaceo, Vmbellam, seu Vmbraculum vocant; quia erectum in latam quandam superficiem definit ad eorum similitudinem, quibus vtimur in itinere ad arcendos solis radios. <110> Fructificant autem quaedam in summis caulibus, vt Caepa, et inter arbores Palma. Quaedam in summis surculis, vt Verbenaca, et inter arbores Rhus. Quaedam a latere, idque vel in ipsis foliorum alis, vt Ficus, vel in opposita foliorum parte, vt Vitis, vel in alis ramusculorum in ipsamet diuisione, vt Tithymalus: quae non possunt simul florere: sed particulatim pro ramusculorum germinatione: Quaedam in ipsis foliis fructificant, vt Ruscus, Chamaedaphne; eorum enim folia neruosa sunt, parum differentia a caulis substantia; idcirco sedem
[23] seminibus impartiri potuerunt: sunt | etiam quorum folia adeo crassa, vt in ramos et caudicem transeant, vt Opontia, in cuius summis foliis tanquam in surculis fructus

16 *germinatione quœdam* impr.; C. corrigit.

108. [The floral protection as criterion for individuality]
If the flowers are associated with individual seeds, but beneath a single common floral protection, as we see in acanaceous plants, we must say that there is a single fruit and not several. But if there is no common protection, even if the flowers and fruits together form a bunch, like in the privet (*Ligustrum*) and the vine, we cannot say that there is only one single fruit, but rather a bunch of fruits. On the other hand, there is one single bunch, which is divided into several fruit implantations from one single provenance of the stem.

109. [Different names for the cluster of fruits]
However, it is not called a cluster in all [plants], but [only] when the fruits consist of a pericarp. In the cereal genus, we call it the ear because it is erect; beard (*Iuba*) when it hangs here and there, as in millet (*Milium*); panicle when it is more compact, as in panicgrass *(Panicum)*; in the ferulaceous genus, we call it umbrella (*Umbella*)[433] or parasol (*Umbraculum*) because once it has opened out wide, it projects a certain surface [of shade], just like those objects used when travelling to block the sun's rays.

110. [Fruiting locations]
Some [plants] bear fruit at the level of the top of their stems like the onion (*Caepa*) and, amongst trees, the date-palm; some rather at the level of the highest scions, like common verbena (*Verbenaca*)[434] and, in trees, the European smoketree (*Rhus*); others on the side, either at the level of the leaf stipules like the fig tree, or in front of the leaves like the vine, or at the level of the twig stipules just at the dividing point like the euphorbia. These [plants] cannot flower at the same time [as they fruit], but [flower] especially before the development of the twigs. Some fruit at the level of the leaves themselves, like the butcher's broom and the Dutch butcher's broom (*Chamaedaphne*): their leaves are fibrous and not of a very different substance to that of the stem; for this reason, they can offer an implantation for seeds. Others | have thick leaves for this [23]
purpose, so that they extend into the branches and the stem, like the cactus (*Opontia*)[435], whose fruits emerge at the tips of the leaves of the scions. Some fruit at the level

433 In contemporary terminology, the word is umbel.

434 In his herbarium, Cesalpino lists two *Verbena:* the one we know as common verbena (*Verbena officinalis* L.) but also gypsywort (*Lycopus europaeus* L.).

435 The genus *Opuntia* refers to cacti known as "racket" or "paddle" because of their flattened form: their flowers and fruit are borne on their top rim. Some are edible, like the prickly pear. The Latin text uses the spelling *Opontia*, but it is the same species, since *Opuntia* gets its name from the ancient Greek city of Opus or Opous (Ὀποῦς, gen. Ὀποῦντος).

nascuntur. Quaedam in radice fructificant fructuum pediculos potius quam cauliculos ferentes, vt Viola, Ciclaminus, Mandragoras, Chamaeleon. <111> Quaedam florent priusquam folia edant et germina, vt Amygdala: fructificant enim in annotino surculo. idem contingit iis, quae in ramis grandioribus et caudice fructificant, et inter herbas quaedam ex iis, quae in radice fructificant, vt Cholchicum, Narcissi quoddam genus, Ciclaminus. Quaecunque autem in caulibus fructificant, et inter arbores, quae in nouellis germinibus, omnia germinant prius, et folia edunt quam flores, sunt quaedam, quae vtroque modo fructificant vt Ficus: quidam enim in annotinis surculis proueniunt, quos grossos vocant, quidam in germinibus nouellis, qui et posterius maturantur; <112> Tempora autem fructificationis sunt pro germinationis tempore, cum ad perfectam aetatem prouenerint, in qua gignere possint sibi similia, Vere autem plurima tum germinant, tum fructificant inter arbores, <113> sed de fructificatione et partibus ad fructus constitutionem datis haec sufficiant in vniuersum; nam singillatim in singulis differentias contemplabimur.

of the root and bear fruit peduncles rather than little stems,[436] like the violet (*Viola*), spring sowbread, mandrake, or the stemless carline thistle (*Chamaeleon*).[437]

111. [The moment of fructification depends on the type of growth]

Some [plants] flower before the leaves and shoots have grown, like the almond tree: they fruit on a one-year-old scion[438]. The same happens with those which fruit from the trunk or from bigger branches. Some of them [those that flower before the leaves and shoots have grown] include the herbs that grow from the root, like the autumn crocus (*Cholchicum*), a certain kind of narcissus (*Narcissus*) and the spring sowbread. All [the plants] which fruit from stems, and amongst trees those which [fruit] on new shoots, without exception develop first, and their leaves grow before the flowers. Some fruit in both ways, like the common fig tree. Indeed, some [figs] are produced on one-year-old scions, which we call summer figs (*grossus)*, and some on new shoots, which also ripen later.[439]

112. [Fructification depends on development]

The stages of fructification depend on the stages of development: when [plants] reach maturity, it is at that point that they can produce things like themselves.[440] In truth, most trees sometimes grow, and sometimes fruit.

113. [Conclusion on the parts dedicated to the protection of the fruit and fructification]

Let this be enough generally speaking on fructification and the parts dedicated to making the fruit. We are now going to examine more closely the differences in specific cases.

---

436 In other words, *pediculi* (little feet) rather than *cauliculi* (little stems). Indeed, the plants mentioned do not bear their fruit on the tips of erect stems, but directly from peduncles from the root. When mature, the fruit of the spring sowbread or the mandrake rest on the ground (see image in the commentary).

437 This plant is probably *Carlina acaulis* L.

438 Almond trees, like other trees, only bear fruit on twigs that are one year old or older.

439 Cesalpino is describing biferous fig trees, which produce two types of figs during the course of the year. The first are harvested in July on the previous year's branches and are known as summer figs; the second fruits, known as autumn figs are borne on the current year's twigs and are harvested in autumn.

440 This is what seeds do when they have developed in suitable conditions. See chapter 9.

CAP. XI.

<114> SUNT porro aliae quaedam partes in plantis nonnullis, vt in animalibus, cornua, pili, et vngues, partim alicuius gratia, partim ex necessitate; Alicuius quidem gratia plantis, quarum caules infirmiores sunt, quam vt se ipsas aut fructus sustinere possint, natura addidit capreolos, seu clauiculos, quibus tanquam manibus vicinas plantas apprehendant, et veluti funiculo circumducto amplexentur, vt earum adminiculo sustentari possint, vt Viti, Cucurbitae, Piso. Oriuntur autem capreoli, vel in ipsis foliorum alis, vel a latere, vel ex aduerso, vel etiam in summis foliis, vt leguminibus quibusdam contingit; Quaedam foliorum pediculis pro capreolis vtuntur, vt Clematis: Hedera aliud genus capreoli sortita est; nam veluti vngues per totum caulem frequentes, aut veluti Centipedum, pedes disponuntur, quibus pertinaciter affigitur arboribus et parietibus. Quod autem capreolorum instrumento quaedam faciunt, aliae suismet germinibus praestant serpentis modo se circumuoluentibus: Nam hoc modo vicina amplexantes scandunt, vt Periploca, et inter herbas Helxine, quasi sensus quidam adiacentis corporis illis videatur inesse, cum repant, donec inueniant, et inuentum apprehendant. <115> Aliis ad arcendas iniurias, aut etiam inferendas tributi sunt aculei; nunc in

12 *capreolorum*] *capreoloruw* impr.; corrigimus.

Chapter 11 [The secondary parts]

114. [Tendrils]

Many plants have other parts as well, just as animals have horns, hair, and nails, partly for a purpose, and partly out of necessity.[441] To those plants whose stems are weaker, for the purpose of supporting themselves or their fruit, nature has added tendrils (*capreolus*) or hooks (*claviculus*): hands, so to speak, with which they can grab neighboring plants and grip onto them, wrap around them like a rope, so that they can hold onto their support, like the vine, the marrow, and the pea. Tendrils originate either at the level of the stipules of the leaves, or on the side, or in front of [the leaves], or even at the level of the outermost edges of the leaves, as happens in some legumes.[442] Some [plants] use the peduncles of leaves like tendrils, like the clematis (*Clematis*).[443] Ivy benefits from another kind of tendril. Indeed, what look like claws[444] (or centipede feet) are arranged at regular intervals along the whole stem, thanks to which they fix themselves solidly to walls and trees. Some [plants] do the same thanks to their tendrils; others are notable for their shoots which twine like a snake.[445] Indeed, in this way, they climb by wrapping themselves around their neighbors, like the silkvine (*Periploca*) and, amongst the herbs, baby's tears (*Helxina*), as if they possessed a kind of sensation (*quasi sensus quiddam*) of the body next to them, since they creep until they find it, and having found it, grab onto it.[446]

115. [Thorns]

Others have thorns, either to protect themselves from harm, or in an act of retaliation. [These are found] sometimes on the stems as in the case of the raspberry, or on the

441 Some parts of the body have a precise function with a purpose; others are the inevitable result of other causes. Cesalpino might be inspired here by Theophrastus, *Historia plantarum* I, 1, 3 (ed. and Fr. transl. Amigues 1988–2006: 1.9–10; ed. and Engl. transl., Hort 1916–1928: 1.5–7). According to Aristotle, *Metaphysica* Δ 5 1015a20–b16 (ed. Jaeger 1957: 92–94), what is necessary is what cannot be otherwise, either for metaphysical reasons or by pure material constraints. Here, Cesalpino is only calling upon this second aspect: its "necessity" is not one of substance, but the inevitability of causal links which happen to generate parts without their own purpose, like moss or spongioles. This is in line with how Aristotle uses necessity to mean efficient causality in nature. See *De partibus animalium* I, 1, 639b11–640a10 (ed. and Fr. transl. Louis 1956: 2–3).

442 Some plants in the Fabaceae (legume) family, such as the tufted/bird vetch (*Vicia cracca*) or the pea (*Pisum sativum*), develop their tendrils at the extremity of their leaflets (see picture in the commentary).

443 The petiole (peduncle) of the clematis leaf indeed serves the purpose of a tendril (see picture in the commentary).

444 These are ivy's crampons.

445 He is referring to twining stems, like those of the bindweed, which have given their name to its family, the *Convolvulaceae*, from the Latin term *convolvere*: to coil or wind.

446 It has not escaped Cesalpino's notice that climbing plants feel their way along adjacent objects, which seems to imply a sense of touch, and therefore sensation. On that idea, and the problems it raises, see the commentary.

caulibus, vt Rubo; nunc in foliis, vt Agrifolio; nunc in fructu, vt Tribulo; nunc sub foliis
[24] tantum, vt Ononi: nunc vbique, vt plu|rimis Acanaceis: Quibusdam surculi ipsi in pungentes aculeos desinunt, vt Iuncus acutus, et Nepa, aliis ipsamet folia aculei sunt, vt Asparagus. <116> Partes autem ob solam necessitatem additae videntur, quae in vetustis arborum caudicibus et ramis innascuntur, vt villi quidam conuestientes pilorum modo, aut ad similitudinem Iubae propendentes, quem Muscum vocant; Peculiaria etiam quaedam sunt, vt in Robore Galla, in Ilice granum puniceum, in Vlmo vesicae quaedam, in Rubo Canino villosi quidam tuberculi, quae spongiolae vocantur, et in aliis multis diuersa, quae tanquam excrementa sunt iudicanda, cum nullius gratia data esse

---

6 *Muscum*] *Mogeum* impr.; C. corrigit.

leaves as in the case of the holly (*Agrifolius*),[447] sometimes on the fruit as in the goat's
head, sometimes beneath the leaves like in the spiny restharrow (*Ononis*), sometimes
all over like in the majority | of Acaenaceous plants. In some [plants] the scions them- [24]
selves detach as perforating thorns, like gorse (*Nepa*)[448] and spiny rush (*Iuncus acu-
tus*);[449] in others, like the asparagus (*Asparagus*), the leaves themselves are thorny.[450]

116. [Accidental parts]

Some parts seem to be added on only by necessity: those which appear on the old trunks and branches of trees, like clumps of sorts which cover them like hair, or which hang down like a beard (*Iuba*), and that we call moss (*Muscus*).[451] Some are particular: the gall[452] in the oak (*Robur*); a red spot[453] in the kermes oak (*Ilex*);[454] a pouch of sorts (*vesica*) in the elm (*Ulmus*); bushy excrescences on the dog rose (*Rubus Caninus*)[455] that we call spongioles[456] (*spongiola*); and various [parts] in many other plants, all of which we must consider as excretions, since they seem to be given for no purpose, even if they do have some use for humans.[457]

447 Cesalpino describes it in Book III, ch. 6. It looks like *Ilex*, to which he compares it.

448 Cesalpino writes about this in Book III, ch. 37: Theodore Gaza translates *nepa* for what Theophrastus calls "scorpion" and what is commonly known as "Christ's thorn" or "crown-of-thorns." Using Cesalpino's herbarium, Caruel (1853: 158) identifies it as *Ulex europaeus* L.: gorse.

449 *Juncus acutus* L. in current binomial nomenclature.

450 The genus *Asparagus* contains a dozen wild species in Europe. Wild asparagus with sharp leaves (*Asparagus acutifolius* L.) which grows on the Mediterranean coast, is thorny. The Cesalpino herbarium specimen referred to as *Asparagus* is identified by Caruel (1853: 139) as an asparagus with narrow leaves (*Asparagus tenuifolius* Lam.), but this species is not thorny.

451 What Cesalpino describes here as the accidental parts of trees are actually lichens or epiphytes or parasites like beard lichens (*Usnea barbata*) (see picture in the commentary).

452 Galls are excrescences (also known as cecidia) formed in reaction to parasites.

453 Kermes oak (*Quercus coccifera*) leaves can develop galls which take the form of red spots (see picture in the commentary).

454 Cesalpino describes it Book II, ch. 2: 32–35. See commentary for more explanations and pictures.

455 Botanists have not resolved the identification of this species. Rabelais 1532 (ed. Esmangart, 1823: 5.281) indicates that the Romans used *Rubus caninus*, the dog rose (*Rosa canina* L.) for what the Greeks call χυνάζ, sometimes also identified as hawthorn after the plant called *cyna* or *cynus* by Pliny, *Naturalis historia* XII, 11 (§22) (ed. and Fr. transl. André et al. 1947–2015: 12.25–26; ed. and Engl. transl. Andrews et al. 1938–1962:4.17).

456 These are "spongiform" growths which do not correspond to what current botanists called spongioles, namely the tips of the rootlets. The description might match the dog rose gall which is red and with a hairy appearance (see picture in the commentary), which would support this identification. The dog rose gall is known as the rose bedeguar gall, or Robin's pincushion (after the woodland sprite Robin Goodfellow) and was used in medicine.

457 Interesting clarification here on the teleology which structures his argument: the purpose to which he refers (and which is "given" by nature) does not include a purpose for humans, only a purpose for the plant. Oak galls are indeed used to make ink, and wood galls used for burr or burl wood which is a precious commodity in woodworking for making veneers.

videantur, licet hominibus vsum aliquem praestent. <117> His autem explicatis sequitur, vt plantas in genera distribuamus.

CAP. XII.

<118> Svnt quidem omnibus recepta quatuor genera plantarum ex totius habitu et vita distincta: Arbor, Frutex, Suffrutex, Herba; <119> ex totius igitur habitu Arbor et Frutex a reliquis differunt: nam corpulentia illorum longe durior grauiorque spectatur, quod lignum vocatur. Reliquorum autem mollior substantia, et laxior est, ideo et diutius viuunt Arbores, Fruticesque, et vt plurimum magnitudine quoque vincunt. Arbor autem a Frutice differt simplicitate; vnico enim caudice apta nata est arbor assurgere, nisi impediatur, eoque habitiori, quam Frutex; Hic autem pluribus, iisque gracilioribus, nisi arte coerceantur; affinis enim eorum natura est: Suffrutex autem et herba inuicem ex viuendi tempore distinguuntur. <120> Suffrutex enim perplures annos viuit, aut radice tantum, vt Ferulacea, et Acanacea, quorum caules singulis annis pereunt, aut etiam caule, vt Ruta, Sampsucus: ideo saepius semen ferunt: ex vitae igitur diuturnitate videntur accedere ad Fruticum naturam magis, minus; recedunt autem ex mollioris

117. [Conclusion on the description of the parts]
The way in which we categorize plants into genera follows from these explanations.[458]

Chapter 12 [The four main genera and their divisions]

118. [The four genera]
In all, there are four genera of plant, quite distinct in their habit (*habitus*) and their form of life (*vita*): trees (*Arbor*), shrubs (*Frutex*), undershrubs (*Suffrutex*) and herbs (*Herba*).

119. [Differences between trees and shrubs]
In general, trees and shrubs differ from the others in their habit: their body (*corpulentia*), which we call wood, appears much harder and heavier. The substance of the others is softer and more supple. It is for this reason, and also because they are bigger, that trees and shrubs live a long time. Trees differ from shrubs by their nature of being a single unit: by virtue of its nature, a tree grows only from a single trunk (unless it is thwarted) and they are bigger than shrubs. The latter has several trunks, which are slenderer (unless one intervenes to restrict their number). Trees and shrubs are indeed related in nature, whereas undershrubs and herbs are distinguished from each other by their life cycle.

120. [Undershrubs]
Indeed, undershrubs live for several years: either as a root, like in Ferulaceous and Acaenaceous plants (whose stems only last a few years); or with the stem as well,[459] as in rues (*Ruta*) and wild marjoram (*Sampsucus*).[460] It is for this reason that they [undershrubs] bear seeds more frequently. They therefore seem closer in nature to shrubs by their longevity, but they distance themselves by the have a more tender nature of their substance.

458 In other words, the classification that follows will take into account all the explanations about the different parts of plants mentioned in the chapters leading up to this one.

459 Undershrubs can live for several years, either surviving as a single root (while their stems die), or as a stem and a root together.

460 The identity of *Sampsuchus* as wild marjoram is not certain, because this is a plant that tends to lose its stems from one year to the next, even though it stays alive (perennial). Nevertheless, Caruel (1853: 128) identified the specimen of *Sampsucum* in Cesalpino's herbarium as this plant (*Origanum vulgare* L.).

substantiae natura. <121> Si quae autem in eo genere, aut etiam in genere herbaceo solidiori corpore constent, et habitiori, arborescentia cognominari solent, vt in genere Maluae, et Tithymalorum, et Ferularum reperiuntur. Aliter autem Suffrutex quoque accipitur, qui surculosus assurgit similitudine quadam Fruticis, quod et inter herbas reperitur, qui et Fruticulus appellari solet. <122> Herbae autem proprie appellantur, quae postquam semen perfecerunt, moriuntur ex toto, vt Fruges et Olera plurima, <123> quamuis alio modo Herbarij accipiant, herbas appellantes, quae in foliis insitas habent vires, vt etiam rustici appellare solent, quarum folia ad iumentorum pabulum veniunt, et ex quibus foenum parant. <124> Alij herbas vocant, quae a radice foliatae exeunt, et caule non caudice semen ferunt: quasi herbis vnicus caulis insit, vt arbori-
[25] bus vnicus caudex; sic enim a Suffruticibus, et Fruticibus distinguuntur; quae exeunt | surculosa, nec a radice foliata. <125> Sunt autem secundum primam significationem herbarum, quaedam annuae, quae intra annum semen absoluunt et vitam, vt Fruges: Quaedam Bienniae, quae non nisi anno secundo semen ferunt, vt Apium; Triennias et-

10 *et*] *vt* impr.; C. corrigit.

121. [Other meanings of "undershrub"]
And if some plants belong to this genus, or to that of herbs, but with a more solid and bigger body,[461] it is common to call them "arborescent", as is observed in the genera of mallows (*Malva*), spurges and giant fennels. But "undershrub" can also be understood differently: the scion which grows in a similar way to a sort of shrub and is found at the level of herbs (*inter herbas*), is also often called little shrub (*Fruticulus*).[462]

122. [Herbs: literal sense]
Strictly speaking, the term "herbs" is used for those plants which, after having produced their seed, die completely, as is the case with many vegetable varieties (*Olus*) and legumes.

123. [Herbalists' definition of "herbs"]
That said, herbalists[463] understand the term differently. They use "herbs" for those plants whose medicinal properties reside in the leaves, as they also usually qualify with "rustic" those whose leaves, with which one prepares hay, are used to feed draught or pack animals (*iumentum*).

124. [Herbs: botanic sense]
Others use the term "herbs" for those plants which grow directly endowed with leaves from the root, and carry the seed by means of a stem (rather than a trunk): herbs have only one stem, just as trees have only one trunk. This is how they distinguish them from undershrubs and shrubs, which grow in a woody state | and which do not have [25]
leaves at the roots.[464]

125. [Corollary of the lifespan of herbs]
According to the first definition of herbs, some are annual, in other words they release – within the year – seed and life, such as what we harvest (*Frux*); others are biennial, in other words they only bear the seed the second year, like marshworts (*Apium*); we have learnt that others still are triennial.

461 Here Cesalpino uses the more common *corpus* instead of *corpulentia* which he has been using in the last few chapters.

462 The way in which Cesalpino has phrased this paragraph suggests that he does not necessarily endorse the use of these terms or these ideas himself.

463 *Herbarii* can mean botanists as well as herbalists, but the use of "differently" in opposition to "strictly speaking" indicates that Cesalpino is referring here to the latter.

464 This seems to be the position of the botanists that Cesalpino favors, as opposed to that of the herbalists. The definitions are moreover those of Theophrastus in his *Historia plantarum* I, 3, 1 (ed. and Fr. transl. Amigues 1988–2006: 1.9–10; ed. and Engl. transl. Hort 1916–1928: 1.23–25).

iam quasdam nouimus: <126> quo autem modo singula haec genera in species distribuenda sint, difficile est videre; cum enim innumerabilis pene sit plantarum multitudo; necesse est, et alia genera multa esse intermedia, sub quibus vltimae species continentur, sed pauca adhuc nota sunt; <127> nam praeter frumenta et legumina, quae communi nomine Fruges vocantur, et praeter Olera vix alia reperiemus manifesta, et haec ab usu potius assumpta, quam a formae similitudine, quam quaerimus: inter Fruges enim Sesamum numerant, et Irionem; quia eorum seminibus vescimur, formis maxime a caeteris diuersa. In Oleribus autem longe plura reperiemus formis distantissima, vt Asparagum, Brassicam, Portulacam, Cicorium, Bulbos, aliaque multa, quae in nulla re alia conueniunt praeter quam in vsu cibario, vel tenera germina, vel folia, vel radices, vel totas plantas nobis assumentibus. <128> Inter suffrutices quoque pauca quaedam genera notantur, vt Coronarium, quod eorum flores, vel folia ob quandam coloris venustatem, vel suauitatem odoris in coronas assumi soleant: cum tamen in caeteris inter se valde distent. Idem in genere aculeato reperiemus, et ferulaceo: praeterquamquod haec a quarundam partium similitudine assumpta sunt; Ferulaceum enim genus ponitur, quod caulem similem Ferulae gerit: aculeatum vero, cuius vel folia, vel fructus, vel caulis aculeis armatur: quamuis huiusmodi inter frutices quo-

126. [The division of genera into species is not easy]
It is difficult to discern how each of these genera ought to be divided into species, because the quantity of plants is almost innumerable. Many other intermediate genera, under which these last species would be categorized, must necessarily exist, but few are known for the moment.

127. [Dietary criteria]
Furthermore, we will discover that cereals and legumes, for which we use the common term "crops", as well as vegetable varieties, are quite[465] different, and that these genera are categorized together more as a result of their use (*ab usu*) than because of their similitude in form, which is what we are studying here. Indeed, included in crops we have sesame (*Sesamum*) and hedge mustard[466] (*Irio*), because we nourish ourselves from their seeds, even though their form is very different from those of other crops. In the vegetable varieties, we find even more major differences in form, for example in asparagus, cabbage (*Brassica*), purslane (*Portulaca*), chicory, onions (*Bulbus*) and many other [species] which conform in no other way but their use in food; [we see that] by looking sometimes at the tender shoots, sometimes at the leaves, sometimes at the roots, sometimes at the whole plants.

128. [Criteria of appearance]
Among the undershrubs too, [some] distinguish only few genera, like that of the crowned (*Coronarium*), whose flowers or leaves often have charming colors or sweet fragrances in their crowns, while in other respects, they differ significantly from each other[467]. We will find the same thing in the genus of thorny plants (*Aculeatum*) and in the ferulaceous genus, with the exception that they are adopted because of the resemblance of some [of their] parts:[468] the ferulaceous genus is defined by its bearing of a stem which is similar to a rod (*ferula*); as for thorny plants, by their leaves, fruit or stem which are armed with thorns; all despite these phenomena being observed just as much in shrubs as in trees or herbs.

---

465 *Vix* usually means "hardly", but as we will see, Cesalpino is using a euphemism here.

466 We know from Pliny that the Romans called this plant *Irio* and the Gauls called it *Vélar.* See his *Historia naturalis* XVIII, 22 (§96) and XXII, 75 (§158) (ed. and Fr. transl. André 1947–2015: 18.89–90 and 22: 77–78).

467 Cesalpino seems here to be criticizing a genus being defined by its use, for example the grouping of undershrubs from which we make garlands. This is a clear allusion to Theophrastus *Historia plantarum*, book VI, subdivided between thorny and non thorny undershrubs, and whose chapters 6, 7 and 8 are dedicated to plants used for crowns. Furthermore, color, taste and fragrance will later be rejected as accidental characteristics which are not suitable markers of genus. See below, §132, as well as chapter 14, §155–156.

468 Cesalpino moves here to subgenera which are not categorized according to the use of the plant, but by morphological characteristics, nevertheless less precise than what he proposes hereafter.

que, et inter arbores, et inter herbas reperiantur. <129> In iis praeterea, quae vires habent medicatas, vnum genus herbarum statuunt, cum tamen et multitudine differentiarum, et totis formis maxime omnium vagum sit. <130> Ex his igitur, quae hucusque tradita sunt, difficilis valde est plantarum cognitio: indistinctis enim generibus, species multis modis confundi necesse est; orta autem est difficultas, quia incertum est, vnde similitudo generum colligenda sit. <131> cum enim duae maxime conspicuae sint plantarum partes radix et germen, ex neutrorum similitudine, et dissimilitudine genera, et species colligi posse videntur; nam si vnum genus earum statuamus, quae radice constant rotunda, vt Rapum, Aristolochia, Cyclaminus, Aron, ab eo genere separabimus, quae maxime conueniunt, vt Napum et Raphanum, quae cum Rapo, et longam Aristolochiam, quae cum rotunda; coniungemus autem distantissima. Cyclamini enim et Rapi distantissima est natura in caeteris omnibus. Similiter autem si vnum genus statuamus eorum, quae radice constat fibrosa, vt Tritici, Ranunculi, Ellebori, in easdem difficultates incidemus: quod et in caeteris differentiis contingit. <132> Si vero caulium

129. [Medical criteria]
Furthermore, in those plants that have medicinal qualities, they conceive of only one genus of herbs, despite the multitude of differences between the plants, and the fact that this genus is the most imprecise of all relative to all the forms.[469]

130. [Consequence: these subgenera are vague]
From this, knowledge of plants is quite difficult from these genera which have been passed on up until this point: because of their imprecision, we cannot but confuse species in several ways. The difficulty comes from the uncertainty of how to determine a genus from a similitude.

131. [Differences in the root cannot determine subgenera]
Although two parts in particular are visible in plants, the root and the shoot,[470] we cannot seem to establish genera and species from the similitudes or dissimilitudes of either of these. Indeed, if we assume one single genus for plants whose root is rounded, like swede (*Rapum*), smearwort, spring sowbread and arum (*Aron*), we would exclude from this genus plants which would absolutely have their place in it, like turnip (*Napum*)[471] and horseradish, which are like swede, and European birthwort (*Aristolochia longa*) which is like the rounded [smearwort] (*Aristolochia rotunda*), whereas we would bring together [species] which are very distant. Indeed, the nature of spring sowbread and swede is very distant in all other points.[472] In the same way, if we assume one single genus for those whose root is fibrous, like wheat, buttercup, and hellebore, we will run into difficulties with respect to their other differences.

469 In other words, either all forms of their parts, or the general form of the plants. The single genus used by herbalists is imprecise with regards to its form, that is to say with regards to its substance in an Aristotelian sense.

470 In chapter 1, Cesalpino distinguishes two essential parts in plants, the root and the stem – rather than the shoot: the terms "stem" and "shoot" seem therefore to be interchangeable, as the following paragraphs show.

471 The term swede (*Rapum*) does not have a precise botanical meaning. It encompasses a whole series of cultivated varieties whose flattened or rounded root we eat, including the swede or rutabaga, kohlrabi, rapini or broccoli rabe (cime di rapa), radish and turnip. Nevertheless, Cesalpino distinguishes between the swede (*Rapum*) and the turnip (*Napum*) which he does not include in the swede family. Many old varieties of turnip are elongated (and not rounded like the majority of turnips sold today). Sachs translates *napum* as "rape" in this passage (Sachs 1875; Eng. transl. Garnsey 1890: 52).

472 The argument is as follows: if we base our categorization purely on the shape of the root, we put turnip and horseradish in a different genus from swede, and the two Aristolochias (European birthwort – the long, and smearwort – the round) in a different genus while these plants are very similar in other aspects. Conversely, we group together spring sowbread and swede simply by virtue of their common round root, whereas they have very little else in common.

differentias contemplemur, vt unum genus eorum ponamus, quorum caules nudi sunt,
[26] ut Iunci, Caepe, Aphacae | inter Cicoracea, Violae: similter connectemus diuersissima, et disiungemus maxime affinia. Positum est genus Ferulaceum ob caulis similitudinem, qui propter leuitatem baculis expetitur praecipue pueris cedendis, ferulam comprehendens Asphodelum, Elleborum candidum Papyrum, et alia huiusmodi in caeteris distantissima. Quod si foliorum caulem uestientium differentias notemus, aut etiam florum, in easdem difficultates incidimus; multa enim sunt genere distantissima, quorum folia maxime similia existunt, ut Polygoni, et Hyperici Erucae, et Sesamoidis, Apij, et Ranunculi, et quae eiusdem generis aliquando foliorum differentia ualde differunt, ut Ranunculorum species, et Lactucarum. Tanto minus florum colores aut figurae, aut alia huiusmodi similitudinem genericam plantarum constituent. Quid enim commune habet Vitis cum Oenanthe, praeter floris similitudinem ? <133> Quod si omnium partium similitudinem quaeramus in generibus constituendis, non magis species ultimas quam genera constituemus. Nam quae similia in omnibus sunt, ut plurimum specie non differunt. Modica enim earumdem differentia non semper speciei diuersitatem facit, cum saepe ob locorum diuersitatem, et culturam, multum immutentur tum folia, tum flores, tum reliquae partes, ut praecipue patet in arboribus. Si enim domesticae serantur semine, ut plurimum syluestres nascuntur, quasi indifferentes specie sint, quae cultu, aut alia ratione fuerint alterata: nam simile ubique simile gignit, secundum

---

5 *Asphodelum*] *Asphodolum* impr.; corrigimus.

132. [Differences in the stem, the leaf or the flower cannot determine subgenera]

If we study differences in the stem, in order to determine a single genus for those plants whose stem is bare, like rush (*Iuncus*), onion, dandelion (*Aphaca*) or, | among [26] the chicoraceous genus, violet, in the same way we would put together very different plants and separate very similar ones. The ferulaceous genus is defined by the similitude of the stem – which, because of its lightness is used in canework for carrying children[473] – and includes the rod of asphodels (*Asphodolus*),[474] the white hellebore (*Elleborum candidum*), papyrus (*Papyrus*) and other plants similarly different from each other in other aspects. And if we take into account the differences in coating [of] the stem of leaves or flowers, we would run into the same difficulties: many plants whose leaves are similar are very far apart in their genus, like knotgrass (*Polygonum*) and perforate St John's wort (*Hypericum*); rocket (*Eruca*) and mignonette (*Sesamoidis*); marshwort and buttercup. Those which belong to the same genus sometimes differ significantly in their leaves, like the species of buttercup and lettuce (*Lactuca*). The colors, shapes and other characteristics of the flowers constitute just as small a generic similitude. Indeed, what do the grapevine and meadowsweet (*Oenanthe*) have in common other than the similitude of their flower?

133. [The make-up of subgenera and species does not depend on the number of similar parts]

But if we do look for a similitude in all parts in order to make up genera, we will not constitute more final species than genera. For in the majority of cases, those [plants] which are similar in all aspects do not differ in species. In these plants, the little dissimilarities do not always constitute a difference in species, since often, because of the difference in location or in the way they have been cultivated, the leaves, flowers or other parts of the plant are altered significantly, as is especially evident in trees. Furthermore, if cultivated trees are reproduced from seed, so that many wild trees grow from them, they will be so to speak of the same species [as their parents] but might look changed.[475] Indeed, everywhere "like begets like",[476] in accordance with

473 In other words, giant fennel stems are used in basketry to make Moses baskets and other baskets. Another interpretation allows a different translation, insofar as the stems of the Giant fennel can reach two meters and harden while drying, it is possible to make from them "canes with which children move."

474 Here Cesalpino used the spelling *asphodolus*, which is probably a typographical error for *asphodelus* as it is a hapax legomenon.

475 The text contains the pleonasm "sown from seed", which seems to emphasize that the cultivated trees that he is referring to are not grafted or reproduced by soboles (vegetative reproduction) but rather from seed. This implies uncontrolled cross-pollination with wild varieties, which potentially adulterates the cultivated variety which is not reproduced identically. Nevertheless, it is still the same species since the varieties are interbred. According to Cesalpino, their differences are therefore adjustable and attributable to the way they are grown or to other reasons, most likely the influence of their environment. On this subject, see also Theophrastus, *De causis plantarum*, especially the whole of Book IV.

naturam, et eiusdem speciei; si uero plurimarum partium similitudo sufficiat generi, multa effugient genus proprium, ut Elleborus niger non fuerit eiusdem generis cum albo; tota enim facie differt: similiter se habet Lactuca syluestris cum domestica, praeterquam quod eo modo inuestigantibus plantarum genera non erit manifestum, quo pacto superiora genera sint constituenda; quae enim in plurimis partibus similia sunt, proxima sunt speciebus ultimis. <134> Quae igitur difficultates contingunt in plantarum tractatione huiusmodi sunt. Idcirco Dioscorides secundum uirium similitudines in medicinis coniunxit, et distribuit. Theofrastus genera assumpsit uulgo nota, quae ab usu magna ex parte accepta sunt, alia a loco ut Aquatilia, Montana, et huiusmodi.

CAP. XIII.

<135> Cvm autem formarum similitudines, et dissimilitudines quaeramus, ex quibus constat Plantarum substantia, non autem eorum, quae accidunt ipsis: accidentia enim posterius innotescunt cognita substantia. Idcirco neque ex facultatibus medicatis, neque ex alia vtendi ratione, neque ex locis, in quibus proueniant, aut aliis huiusmodi genera eorum, et species sunt constituendae: Haec enim omnia accidentia sunt. sed
[27] sub|stantiae ratio forte incognita est, incognitis differentiis vltimis, vt multi putant, idcirco oportere per accidentia circumscribere. At contra hanc sententiam disputatum

nature and identical by species. If the similitudes between a large number of parts were enough to establish a genus, many [species] would break away from their own genus, for example the black hellebore (*Elleborus nigrum*) would not be in the same genus as the white hellebore, because the appearance of these two plants differs – the same being true of wild lettuce (*Lactuca sylvestris*) in comparison with the cultivated variety (*Lactuca domestica*) – so that having studied the genera of plants in this way, the criterion for selecting the higher genus is not obvious. So, plants which are similar in the majority of their parts are close in terms of their ultimate species.[477]

134. [Conclusion on the classification of genera and subgenera]
These are therefore the difficulties related to the treatment of plants. This explains why Dioscorides groups together and categorize plants according to the similitudes of their medicinal properties. Theophrastus accepts the commonly accepted genera, largely derived from practical usage, and [brings in] others based on their location, like "aquatic" plants, "alpine" plants, etc.

Chapter 13 [Criteria for defining subgenera and species]

135. [Accidental properties cannot be used to define the species]
Since we are investigating similitudes and dissimilitudes in the forms in which the substance of plants consists, and not the forms which are there by accident (indeed, accidents appear to us once we know the substance), their genera and species must not be based on their medicinal properties, nor on any other usage, nor according to where they come from, nor on any other similar criteria: these are simply accidental. Nevertheless, | perhaps the definition of the substance remains unknown as long as [27] the ultimate differences are unknown,[478] which is why many people think that it would be enough to establish it through accidents.[479] However, I have provided numerous arguments in opposition to this conclusion in *Quaestiones peripateticae*.[480]

476 The quotation marks are added to emphasize the proverbial expression.
477 Defining genera from resemblances from the beginning leads to undesirable consequences, and it is therefore not a good default method, except perhaps as a last resort, and used to distinguish one final species from another. However, one should only take into account a similitude in a large number of parts if all other more precise categorization criteria have been followed for classifying plants in their higher genus.
478 This is classic Aristotelian doctrine: the form or substance of a thing is identified by its last *differentia*. See *Metaphysica* Z 12, 1038a19–23 (ed. Jaeger 1957: 155).
479 The grammar of this sentence is incorrect (neither *ut* nor *idcirco* take the infinitive), and so we must make an interpretation. The argument seems, however, quite clear: some people interpret Aristotle's doctrine by confusing specific last differences with accidents, and so define genera by what are in fact accidents.
480 Cesalpino 1571: I, 2 and 5–6.

est abunde in quaestionibus Peripateticis. <136> Alij differentias secundum formam ex anima tantum colligi oportere putantes, coguntur fateri omnes plantas vnius speciei esse, cum vnicam animae partem, quae vegetatiua appellatur, sortitae sint omnes. At ostensum illud quoque est, differentias formam constituentes etiam ex materia, quae illius gratia data est, colligi oportere, si igitur in plantis indifferentes essent partes ad operationes secundum illam animae partem praestandas, vna esset omnium plantarum species. Quoniam vero videmus multis modis differre, idcirco in multas species, et genera distribui necesse est. <137> Cum igitur omnis substantiae ratio a fine petatur; propter illum enim susbtantiae quoque sunt, quae illius gratia habentur, videndum est in plantis, quae similitudo et dissimilitudo in iis fuerit, quae primi animae operis gratia data sunt, deinde quae secundi, et siquae alia sequantur deinceps. Est autem primum uegetatiui opus, quod omnibus viuentibus inest alimenti attractio, quo nutriantur et crescant. Partes autem huius gratia datae sunt radix et germen: ex horum igitur differentiis prima genera constituenda sunt: vt quorum radices et germina habitiori

3 *sortitae*] *sorbitae* impr.; C. corrigit. **10** *quae*[1]] *que* impr.; corrigimus.

136. [The faculties of the soul are not enough to define the species]

Others think it is appropriate to infer [specific] differences from the form derived from [the faculties of] the soul and therefore consider that all plants belong to the same species, since they have all received the same part of the soul,[481] known as the vegetative part.[482] But we can show that the differences that make up form can be obtained from matter, which [plants] are endowed with for this purpose. But if the parts of plants were not differentiated by the specific purpose they fulfil relating to this part of the soul, all plants would be of the same species.[483] But since we observe in reality that they differ in many ways, distributing them into several genera and species is necessary.

137. [Constitution of the main genera from the first operation of the soul based on the criterion of the hardness of the roots and shoots]

We seek the reason for all substance from its purpose, notably because substances are what one has in view of this [purpose] (*illius gratia*).[484] In plants, we must therefore observe what their similitudes and dissimilitudes are: those made for the first operation (*opus*) of the soul, then those made for the second, then those made for subsequent operations.[485] The first operation of the vegetative [soul] – namely, as in all living things, the acquisition (*attractio*) of food – is that which allows it to feed and grow. The root and the shoot are the parts that are dedicated to this; it is therefore based on these parts that the primary genera must be established, so that those whose roots and

481 For Aristotle, the parts and the faculties of the soul are synonyms. See, for example, *De anima* I, 1, 402 b10–14 and II, 4, 415a14–22 (ed. Ross 1961: 2, 33). Cesalpino means here that plants only have the vegetative (or nutritive) faculty.

482 Some Aristotelians apply the same classification into genera to plants as the one used for animals, which is founded on the sense-faculties of the soul from which they derive their defining form. However, this would lead us to the unwelcome conclusion that there is only one genus for plants, since they all have exactly the same faculties because of the homogeneity of their vegetative soul. Indeed, the vegetative soul would not allow for the diversity of specific expressions that enables the sensitive and intellective parts of the soul, which are more heterogeneous in the animal kingdom. See our commentary.

483 It is not solely a question of the self-evidence of common sense, as it is not enough of an objection for the Aristotelian definition of species: Aristotle indeed only recognizes one species of human, on the basis of the homogeneity of the intellective part of their soul, see, for example, the *Ethica Nicomachea* X 7, 1178a5–9 (ed. Bywater 1894: 214–215). Cesalpino will return to this Aristotelian argument and develop it in §142.

484 This is a powerful way of defining substance: it boils down to what a being has at its disposal to accomplish a single purpose or multiple purposes. For Aristotle, the substance of animal species must be defined by their vital functions. See Book 1 of *De partibus animalium*, passim, particularly 639a11–b9, 641a14–31, and 641b13–642b4 (ed. and Fr. transl. Louis 1956: 2–3, 7, 9–10, respectively; ed. and Engl. transl. Peck 1943: 53–57, 69–71, 73–79, respectively), and *De generatione animalium* I, 1, 715a8–9 (ed. and Fr. transl. Louis 1961: 1; ed. and Engl. transl. Peck 1943: 3).

485 We will see that Cesalpino states below that there are only two main tasks of the soul: he is probably referring here to purposes that are subordinate to the two main ones, such as the protection of the fruit that is offered by the leaves.

substantia, et duriori constant, quod lignum vocatur, arbores sunt et frutices. Quorum vero tenuior et mollior substantia est suffrutices sunt et herbae: Petitur enim haec differentia ex natura similarium partium totas plantas constituentium. <138> Similiter altera diuisione addita vtraque genera subdiuidi possunt: vt quorum simplex germen assurgit, arbores sunt, et herbae, quorum vero multiplex, Frutices sunt et Suffrutices: sumptis ex numero differentiis. sed quoniam Suffruticis et herbae nomen vario modo accipitur, vt superius notauimus, clarius agemus, si altera diuisione neglecta, duo tantum genera constituamus cum arboribus coniungentes Frutices, et cum Suffruticibus herbas. <139> Praeterea ex propriis radicum differentiis, et caulium singula genera summam diuisionem patiuntur. Vt si radix erecta vnica vel plures, vel obliqua, vel geniculata, vel fibrosa, vel rotunda, alba, russa, crocea et caetera huiusmodi; similiter si caulis rectus, aut obliquus, humi stratus, aliena scandens, concauus, geniculatus, aculeatus, asper, leuis, rotundus, quadratus, ex quibus omnibus et huiusmodi, abditis etiam foliorum differentiis suprema genera distinguuntur: quatenus plantarum partes ad primi operis functionem datae sunt: sed ex his nequaquam alia genera colliguntur, cum hoc ad partes secundi operis spectet. <140> Secundum autem vegetatiui opus est generare sibi simile, quod et perfectione prius est, cuius gratia dati sunt fructus et partes ad fructificationem facientes; cum igitur id non omnibus insit, sed perfectiori-
[28] bus, pro fructificationis similitudine, et dissimilitudine poste|riora genera, tum in

5 *et*[1]] *vt* impr.; C. corrigit.

shoots are made of a harder and more solid substance, called wood, are trees and shrubs.[486] On the other hand, those whose substance is finer and softer are undershrubs and herbs. This difference [in substance] is a matter of the nature of the constituent parts of all plants.

138. [Second criterion: number of shoots]
In the same way, each of these two genera[487] can be subdivided by the addition of another division: those which grow with a single shoot are trees or herbs, whereas those whose stems are multiple are shrubs or undershrubs – the differences relate to the number. But insofar as the name undershrub or herb is understood in various ways, as we have already mentioned, if this second division[488] were ignored, we would clearly come to establish two main genera, bringing together shrubs with trees and undershrubs with herbs.

139. [First level of subgenera established from roots and shoots]
Furthermore, taking account of the individual differences in roots and shoots, each genus can undergo a first (*summa*) division. The root grows singly, or multiple, or obliquely, or folded, or fibrous, or curved, or white, or red, or yellow, etc.; in the same way the stem is straight, or oblique, it creeps along the ground or climbs on other things; it is hollow, bent, thorny, rough, smooth, rounded, square, or composed of all these attributes; etc. On the other hand, the supreme genera differentiate themselves by taking no account of the differences in leaves in favor of the parts of plants dedicated to the functioning of their first operation [nutrition and growth]. From the leaves we do not arrive at all at other genera, because they are related to the parts [dedicated] to the second operation.[489]

140. [Second level of subgenera related to the second operation]
Furthermore, the second task of the vegetative [soul] is to reproduce identically, for which the fruit and the reproductive parts of the plant are designed. Since only the most perfect plants have this, rather than all plants, it is through the similitudes and dissimilitudes in fructification that the secondary genera should be established, | both [28]
in the case of trees and also [in those whose] matter [is] closer to the ground.

486 *Substantia* clearly refers here both to essence and matter (in this case wood), since the material aspect is included in what constitutes the substance of the plant, as Cesalpino has just stated.

487 Indeed, at this stage, Cesalpino is only dealing with two genera: plants with a hard substance and those with a soft one.

488 Literally "the other division", but what follows indicates that he is referring to this division.

489 The principle and first divisions of the supreme genera must be based first on the differences in the stems and roots, because these are the organs that are of prime importance to the nutritive and growth faculties of the soul. Leaves are not relevant here, because their purpose is limited to the reproductive process.

genere arboreo, tum in humiliori materia constituenda erunt: <141> Nihil autem refert siue nominata sint genera siue innominata; non enim omnibus nomina sunt imposita, sed iis magis, quae vtilitatem aliquam conspicuam hominibus afferunt, vt frumenta, legumina; <142> quod si tertium aliquod esset vegetatiui opus, ex illo similiter, et ex partibus illi destinatis, tertia colligenda esset partitio, qua superiora genera in alia diuiderentur. Sed quoniam duobus praedictis absoluitur plantarum operatio, idcirco in illis tantum versabitur generum collectio et partitio. Et merito ex modo fructificandi multa emerserunt plantarum genera: in nullis enim aliis partibus tantam organorum multitudinem et distinctionem natura molita est, quanta in fructibus condendis spectatur. <143> Quemadmodum enim animalia ex sensuum, aut motuum instrumentis plerasque et praecipuas differentias sortita sunt: vltra enim sensum et motum alias operationes non habent: sic plantae in fructificatione, tamquam ultima perfectione admirabilem uarietatem ostendunt. Si similiter animae pars intellectiua instrumentis uteretur: hominum genus ex illorum differentiis in multas species distributum esset. Quoniam uero intellectus, per quem homo est, nullo corporeo utitur instrumento, idcirco natura unicam speciem condidit humanam; et si enim quidam Aethyopes sint, quidam Gigantes, quidam monstruosa aliqua forma, omnia accidentia sunt; non enim qua homines sunt differunt, secundum partem scilicet intellectui seruientem: idem in quibusdam plantis inspicere licet, quae specie differre videntur, quarum tamen differentiae accidentia sunt, sed quo pacto distinguenda haec sint, ab his quae substantiam constituunt, inferius patebit. <144> Enitamur igitur ex propriis, quae fructificationis gratia data sunt, plantarum genera inuestigare, tum in arboribus tum in fruticibus, et reliquis humilioribus.

CAP. XIIII.

<145> Plantarum quaedam nullum omnino semen ferunt, quippe quae cum imperfectissimae sint, ex putredine tantum ortum ducant: idcirco eas nutriri tantum et augeri contingit: generare autem sibi simile nequeunt tanquam media inter plantas et inani-

141. [Digression: not all possible subgenera have a specific name]
Nothing prevents genera sometimes being endowed with a name and sometimes not: not all have names attributed to them; when they do, it is usually because they have some sort of obvious use for humans, like cereals and legumes.

142. [There are only two levels of division into subgenera]
If there were a third operation of the vegetative [soul], the same would be true for that and the parts dedicated to it, and we would have needed to include a third separation, by which the preceding genera would be subdivided into new ones. But plant activity is limited to the two operations that we have mentioned, which is why the number of cycles of grouping and separation of genera is the same. And indeed, several types of plants are fittingly revealed through their mode of fructification: nature has endeavored to produce as much diversity and difference in no other part than that which we see in the development of the fruit.

143. [Species are solely defined by their essential operations]
Indeed, just as animals are classified according to the numerous and significant differences in their sensory and motor organs, because they have no other activities but sensation and movement, so plants reveal an admirable variety in fructification, which is like their ultimate accomplishment. If, in the same way, the intellective part of the soul were to use instruments, the human genus would be divided into several species according to the differences between these instruments. But the intellect, by which humans are defined, does not use any bodily instrument, which is why nature has established only one human species. Even if there are Ethiopians, giants, or deformed monsters, these are only accidents: they do not differ as humans, that is, according to the part which uses the intellect. Likewise, we can observe that some plants seem to differ by species, when in fact their differences are accidental; at a later stage we will discuss the criterion (*pactum*) we use to distinguish these differences from those which constitute the substance of the plant.

144. [Transition to the next chapter]
From the properties dedicated to fructification, let us now try to lead our investigation on plant genera, both trees as well as shrubs and other plants closer to the ground.

Chapter 14 [Subdivisions based on the reproductive operation]

145. [Plants without reproductive operation]
Some plants do not bear seeds at all, which makes them imperfect; they draw their origin from putrefaction, which is why they feed and grow to such an extent [as compensation]. They are therefore unable to reproduce identically, being mid-way between plants and inanimate beings (in the same way as zoophytes are between ani-

mata, quemadmodum Zoophyta, inter animalia et plantas, ut Fungorum genus; Len-
ticula palustris, Lichenes, Frutices multi marini. <146> Quaedam moliri quidem semen
uidentur, sed imperfectum efficiunt ob propriam naturam, ut inter animalia Mulus;
sunt enim ueluti aliarum plantarum aborsus aut morbi, ut multa in genere frumento-
rum, quibus spica inanis et fere Iuncorum genera, et Testiculorum, Orobanchae, Hy-
[29] pocistis: nam in his | pro semine puluisculus, aut muccus conceptaculis continetur.
<147> Quamuis autem inter perfectiores, quaedam sint steriles, non sunt in hoc genere
ponendae: quoniam imperfectio in illis non ex natura speciei contingit, sed indiuidui.
<148> Quaedam aliquid ferunt, quod proportione respondet semini; nam per id uisae
sunt propagari; est autem lanugo quaedam foliis insidens, caule autem et flore et
semine carent, ut Filix, Adiantum, Phillitis. <149> Quaedam tandem semen perfectum
ferunt, quod genus cum multam habeat latitudinem, idem primo loco partiamur;
continet enim plantas perfectiores. Cum ad organorum constitutionem tria maxime
faciant, scilicet, partium numerus, situs, et figura, <150> (magnitudo enim non uidetur
speciem organi immutare, nisi simul figuram immutet: Solutio autem continui, aut
unio ad numerum pertinent: durities, mollities, color, et reliquae qualitates ad simila-
res partes referuntur) natura secundum illorum differentias in fructibus condendis

11 *Filix, Adiantum Phillitis* impr.; corrigimus.

mals and plants), for example the genus of mushrooms (*Fungus*), the ivy-leaved duckweed (*Lenticula palustris*), lichens (*Lichen*) and many marine shrubs (*Frutex mari*).[490]

146. [Plants with faulty seed]
Some plants seem to try to produce a kind of seed, but because of their own nature, the result is imperfect, like the mule in the animal world. Indeed, these plants come out of a kind of abortion or disease in other plants. This frequently happens in the cereal genus, where the ear is empty, and often [in] the genera of rushes, orchids (*Testiculus*), broomrapes (*Orobancha*) and cytinus (*Hypocistis*). Indeed, they contain, in [a kind of] receptacle, | a fine dust or a mucus by way of seed.[491] [29]

147. [Sterile plants do not form a separate species]
Although among more perfect plants some are sterile, these must not be placed in this genus, to the extent that, in their case, the imperfection relates not to the nature of the species but to the individual.

148. [Plants endowed with a seed equivalent]
Others [only] bear something that corresponds analogically to a seed: we see that they can propagate from this. It is a sort of down that is found at the level of the leaves. On the other hand, these plants are deprived of a stem, flower, or seed. They include the royal fern (*Filix*), southern maidenhair fern and hart's-tongue fern (*Phyllitis*).[492]

149. [Three criteria for subdividing plants with a perfect reproduction process]
Finally, other [plants] carry a perfect seed, but this genus is extensive; let us therefore divide it in the first place, because it contains more perfect plants. Three main properties contribute to the constitution of the organs, namely the number of parts, their position, and their shape.

150. [These criteria are exhaustive]
Indeed, size does not seem to alter the species of organ without immediately altering its shape; the distention or the cohesion of the adjacent parts fall under the number; the firmness, the softness, the color and other qualities boil down in a similar way to the [shape of the] parts.[493] Consequently, nature has played with these differences by

490 He is probably referring to large corals or perhaps large algae.
491 The seeds of orchids, broomrapes and cytinus are indeed imperceptible to the naked eye; we can only see them in the form of "a fine dust."
492 These are species of ferns which do not therefore produce flowers; their leaves have a "downy" sporangium beneath them (see image in the commentary) which is probably what Cesalpino is describing.
493 The Latin text and its punctuation are not clear here. We understand that if a part is more or less united to another part, they are rather one or two parts, which is a matter of number. The properties mentioned afterwards seem to belong to shape, mentioned at the beginning of the sentence, but this is not explicit. Moreover, the position is not discussed.

multis modis lusit, ex quibus uaria plantarum genera constituta sunt. <151> Nam primum, cum flos et floris tegmen extimum sit fructus inuolucrum, aut sub vno flore vnum semen condidit, vt Amygdala, vel vnum seminis receptaculum, vt Rosa, aut duo semina vt Ferulacea, vel duo seminum receptacula, vt Nasturtium, aut terna eadem vt Tithymalorum genus terna semina, Bulbacea terna conceptacula: aut quaterna, vt Marrubium quaterna semina, Siler quaterna conceptacula, aut plura vt Cicoracea, et Acanacea plura semina, Pinus plura seminis conceptacula: <152> Deinde seminis situs in suo conceptaculo, vel in sede, aut eo modo est, vt eius cor exterius spectet vel interius. Similiter de situ floris considerandum: nam aut exterius summis fructibus insidet, aut inferius circa sedem fructus exoritur. <153> In omnibus autem dictis considerandа est vltimo figura tum seminum, tum conceptaculorum, tum florum: Amplius fructuum sedes, mollities, durities et caetera huiusmodi, ex quibus propinquiora subinde genera constituuntur. Haec vero eadem secundum proportionem reperiuntur tum in genere arboreo, tum in humilioribus. <154> Si quae autem his differentiae accedant, ex foliis caulibus, radicibus, aut aliis partibus, quae ad fructus constitutionem non faciunt, quodammodo per accidens erunt, quodammodo autem per se, quatenus fructus gratia datae sunt, vt folia ad vmbram praestandam, caules ad fructus ferendos, radices ad alimentum subministrandum; idcirco ex illis differentiae, cum iis, quae ex fructificatione sumuntur vsque ad species vltimas genera contrahunt. <155> Quaecun-

5 *terna conceptacula aut quaterna* impr.; C. corrigit.

incorporating them in the fruit in many ways, from which the different genera of plants were constituted.

151. [First criteria: number of parts]

Firstly, as the flower and what protects it on the outside are the husk of the fruit, beneath the flower is situated either one single seed – as with the almond –, or one single seed container (*receptaculum*)[494] – as with the rose –, or two seeds – as with Ferulaceous plants –, or two seed containers – as with cress (*Nasturtium*) –, or three of them: three seeds – as with euphorbias, – or three receptacles – as with bulbaceous plants –, or four: four seeds – as with horehound –, or four receptacles – as with the spindle (*Siler*) –, or several: several seeds – as with chicoraceous and acaenaceous plants –, or several seed receptacles – as with pine.

152. [Second criteria: position]

Next, the position of the seed is found at the level of its receptacle: either at the level of its implantation (in other words so that the heart of the seed is orientated outwards), or inside it. In the same way, we must look at the position of the flower: either it is situated on the outside of the tops of the fruit, or it grows lower, around the fruit implantation.[495]

153. [Third criteria: shape]

And among all these considerations, we must lastly turn to the shape, sometimes that of the seeds, sometimes that of the receptacles, sometimes that of the flowers, and more generally at the implantation of the fruit, its softness, hardness and other [qualities] from which the corresponding genera are then established. In reality we can observe all of this analogically, as much in the genus of trees as in genera closer to the ground.

154. [Subdivision in species according to the secondary reproductive organs]

But if those differences arise in these genera from the leaves, stems, roots, or other parts of the plant that do not contribute to the creation of the fruit, they are in one way *per accidens* and in another way *per se*, in terms of how they contribute to the fruit: the leaves serve to give it shade, the stem to carry it, the roots to provide it with nutrition. As a consequence, the differences obtained from these [parts], alongside those which come from the fructification, differentiate the genera as far as ultimate species.

494 *Receptaculum* (translated as container) is only used in this sentence, with a less precise meaning than *conceptaculum* (translated as receptacle).

495 This corresponds to the distinction between species with an inferior ovary or with a superior ovary (see illustration in the commentary).

que autem neque ad totius plantae, neque ad fructus constitutionem faciunt, vt colores, odores, sapores, et alia huiusmodi, accidentia sunt; Ideo saepe aut ex cultu, aut locorum, aut coeli diuersitate variantur. Quae autem per se sunt, vbique eodem modo spectantur. <156> Quaedam tamen sequuntur naturam specificam, vt plantarum vires in | [30] medicinis, sapores, et aliae proprietates, quas Medici in primis considerant, quae etsi aliquando per se insint, non tamen differentiae sunt constituentes earum substantiam, quas hoc loco quaerimus. <157> Incipiamus igitur ab arboribus cum quibus ob affinitatem Frutices quoque coniungemus secundum rationem praedictam in genera et species partiendo; Deinde ad Suffrutices et herbas accedemus.

Finis Primi Libri.

155. [Other differences are accidents]
As for those differences which do not contribute to the plant as a whole nor to the make-up of the fruit, like colors, scents, vapors, and other things of this type, these are accidents. Furthermore, they often vary depending on the method of cultivation, the location or climate differences. In contrast, *per se* differences manifest themselves everywhere in the same way.

156. [Medicinal properties are only *per se* accidents]
Finally, the differences which are a result of specific nature, like the medicinal proper-
ties | of plants, their taste and other attributes in which physicians are primarily inter- [30]
ested, are however not constituent to the substance, even if they are somehow present *per se*.[496]

157. [Outline of the subsequent books]
We will therefore begin by dealing with trees and shrubs together, on account of their affinity, while dividing them into genera and species according to our established logic; then we will turn to undershrubs and herbs.[497]

End of the first book

496 For Aristotle, some accidents are "*per se*": they result from the definition (and therefore the substance) of the thing without belonging to this definition. See *Metaphysica* Δ 30, 1025a31–32 (ed. Jaeger 1957: 120); see also *Analytica Posteriora* I, 4, 73a35–b24 and I, 22, 83b17–20 (ed. Ross 1949: 119–120 and 144).

497 This outline is indeed followed: books II and III are devoted to trees and shrubs; books IV to XVI to undershrubs and herbs.

Fr. Dionyſius Conſlacciarius facultatem imprimendi concedit. Die 27. Septembris. MDLXXXI.

FLORENTIAE,

Excudebat Georgius Mareſcottus.

MDLXXXIII.

Figure 9: Last page of *De plantis libri XVI*, 1583.

Part III: **Commentary**

https://doi.org/10.1515/9783111001104-003

Introduction to the Commentary

The following pages address the main philosophical and botanical theses of Book I, chapter by chapter. Identification of species and the translation of their names are discussed in the footnotes to the translation.

In our commentary, we consider Cesalpino's scientific theses against the background of the knowledge available in his time, with the occasional inclusion of current positions. The aim is not to propose a historicist reading of *De plantis* by arguing that it contains the first steps towards the resolution of biological problems. Nor is this commentary intended to be positivist, and to assess Cesalpino's work in the light of current scientific knowledge. Rather, we intend to provide an external viewpoint, and insights into the development of the history of botany and medicine as we understand them today. References to current biological explanations highlight the relevance or innovative character of Cesalpino's theses, contrasted to other ideas without empirical basis. This reading also separates more clearly philosophical ideas from theses based on observation, although they often intermingle in Cesalpino's work. Finally, comparisons with current botany allow the 21st-century scientist or amateur better to grasp the realities conveyed by Cesalpino through different vocabulary or unfamiliar methods. Illustrations are included in the commentary for the same reason.

Commentary of the Fourteen Chapters of Book 1

Chapter 1. The Properties and Parts of Plants (§1–14)

Cesalpino first focuses on the life principle of plants and on their ability to feed themselves; he identifies this vital principle with the soul of the plant. He follows Aristotelian metaphysics as he compares the faculties of animals and plants, and the medieval hierarchical conception of the chain of beings as he posits from the outset the ontological deficiency of plants compared to animals. He considers the lack of sensations of plants as self-evident, although this will prove problematic in his analysis of plant nutrition and of how plants select their food in chapter 2, and in his attempt to account for the motions of vines in chapter 11. This will be developed further below.

The "deficiency" of plants lies in the undifferentiated nature of their parts compared to the parts of animals; as a result, it is difficult to explain plant nutrition, since the various organs of the digestive system, and their respective functions, cannot be identified in plants as they can be in animals. Structural analogies between the plant and animal kingdoms are therefore limited to the superficial resemblance of plant roots and animal entrails. Cesalpino is aware of this important limitation (§4), which seems to prevent the transposition of animal physiology and nutrition to plant physiology and nutrition. Cesalpino's philosophical considerations on this issue rely much more on general functions than on physiological functions *stricto sensu*. The main function of the soul – the self-preservation of the living being – manifests itself in two manners: through nutrition (and the resulting growth), at the level of the organism, and through reproduction, at the level of the species and of its perpetuation. Cesalpino identifies in plants two specific parts performing these two functions: the root, which absorbs food, and the stem, which bears the fruit.

Because Cesalpino adopts a predominantly deductive reasoning, like his predecessors and successors until the end of the 18th century, he has difficulty understanding the real nutritional role of leaves:[498] since the stem bears the fruit, its function seems to be reproductive only; therefore the leaves, attached to the stem, must also play a role, albeit secondary, in reproduction (cf. §31) – not in nutrition. On this point, Cesalpino follows Aristotle and Theophrastus who suggest that the leaves' function is to provide shade for the fruit and to prevent it from being burnt by the sun.[499] The Aristotelian analogy between plant roots and animal mouths,[500] and the more modern version of

498 The nutritive role of leaves through the process of photosynthesis, which converts carbon dioxide from the air into sugars from water and solar energy, was not established until 1796 by Jan Ingenhousz (1730–1799).

499 Arber 1950: 28.

500 Aristotle, *De partibus animalium* IV, 4, 678a3–16 (ed. and Fr. transl. Louis 1956: 111; ed. and Engl. transl. Peck 1945: 313–315). This passage is quoted in our section on Analogy in Part 1.

roots as intestinal veins, endorsed by Cesalpino, are misleading.[501] Hierarchical thinking and zoomorphism lead Cesalpino to see the root as the upper part of plants, and the stem as the lower part, and to deduce, from this inferiority of the stem, that it had an excretory function, by analogy with the lower part of animals. This deduction is unsupported by any observation, and indeed Cesalpino cautiously writes that the stem's role is to evacuate excrement, "if there is any" (§6).[502] The analogy is thus not so much asserted than proposed for heuristic purposes.

Cesalpino then addresses the delicate question of the seat of the soul. For Aristotle, the soul of animals is located in a central part, the heart, from which it acts on the other parts of the body; the soul is not spread through all parts of the body or distributed in several singular parts. Cesalpino refers to his *Quaestiones peripateticae*, in which he addresses this issue more thoroughly and shows that it is not necessary for the soul to be present everywhere, provided that it is present in the heart.[503]

> Si igitur animae partes hoc modo se habent inter se, vnum esse oportet omnium principium non plura: idque aut totum corpus indistincte, ut uidetur in ijs quae modicam habent organorum distinctionem, vt plantae, et insecta quaecunque diuisa viuunt: aut corporis aliquam particulam vt ijs accidit, quae perfectiora sunt: Esse autem huiusmodi Cor in ijs quae sanguine praedita sunt, patet. Nam hoc primum oritur cernitur enim palpitare statim in prima foetus generatione. 3 de Part. Anim. cap. 4. Vltimum autem interit, quasi eo non existente, reliquae partes non sint, nisi vt manus mortua, aut oculus mortuus: viuunt enim omnes quatenus adnatae sunt Cordi.

> If the parts of the soul are thus related, it is fitting that it be one principle and not many: it could be either the whole body without distinction (as is apparent in things that have little distinction between organs, like plants and all insects that can survive once divided), or some bodily part in particular (as happens for things that are more perfect). But it is clear that such is the case of the Heart for those that are endowed with blood. For it the first to come into existence, since it is perceived to palpitate even in the first generation of the fetus (*Parts of Animals* III, 4). It is the last to perish, as if when it does not exist, the other parts could not be anything else than a dead hand or a dead eye. For they all live as far as they are linked to a Heart.[504]

However, here too, the metaphysical presupposition that the nutritive function is located in the root, and the reproductive function in the stem, is problematic for the interpretation of plant life, since stems can take root in the earth, and roots can produce aerial stems in the form of shoots. This seems to indicate that the soul is not only found in the plant as an organism, but also in each of its parts, although Cesalpino acknowl-

501 Delaporte 2011: 23, 30 on Cesalpino; see 40–100 on early modern theories of plant nutrition in general.

502 Some early modern botanists thought that the nectar of flowers was excrement (Dughi 1957: 163; Delaporte 2011: 126–128). This idea might fit with Cesalpino's argument here, but he does not endorse this idea about nectar explicitly.

503 Cesalpino 1571: I, 7; V, 3; V, 7.

504 Cesalpino 1571: V, 3: 102c–d.

edges that this is not applicable to all plants (he gives the example of conifers), nor to all circumstances (§10). Because of the variety observed in the different species of plants, it is impossible to formulate a valid, universal thesis for the location of the soul of plants (§11). Although Cesalpino initially follows Aristotle in locating the heart of plants at the level of the root collar, where the root and stem meet (§13), he argues in chapter 5 that the heart of plants, unlike that of animals, has no precise location, but "is found distributed, so to speak, in all parts" (§48: *veluti in omnes partes distributum est*). The status of the plant heart, to which Cesalpino returns in chapter 4, is therefore not very clear. Is it the physical location of a vital organ identified with the root collar or contained within it? Is it rather a substance or a vital humor that emanates from it? Is it a vital, or even metaphysical, principle concentrated in specific parts? The treatise lacks clarity and coherence on this subject.

After these difficult metaphysical considerations, Cesalpino turns to anatomical descriptions. He distinguishes two parts in the root: external (the bark), and internal (the body). He also identifies three distinct layers in the stem: the bark (*cortex*),[505] the intermediate body, which is the wood in trees (*lignum*), and the pith (*medulla*), which is the innermost part (§13) (figure 10).

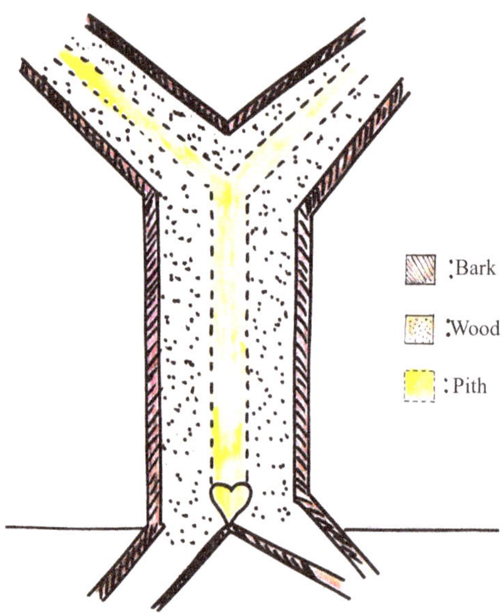

Figure 10: Illustration of Cesalpino's tree anatomy.

505 In §32, he distinguishes the outer part of the bark from the inner part, the bast, which is different in nature.

These observations match current botanical knowledge insofar as the roots generally do not contain pith, but Cesalpino's deductive reasoning, founded upon an *a priori* argument, results in an erroneous conclusion. Cesalpino considers, in principle and by analogy with animals, that the vital principle of plants resides in their innermost parts; thus, the life principle and the reproduction of plants must originate in the most central part of the stem. He calls it "pith", and notes that the Ancients called it "brain", "matrix", or "heart" (§14).

The current botanical term for the most central part of a tree's trunk is "heartwood", which may itself contain the pith in its most central part. And yet, despite its name, this part consists solely of residual dead wood, and is no longer involved in the growth, reproduction, or physiology of the organism. Contrary to the animal model, the living part of trees, and of plants in general, is found under the bark, in the intermediate wood layer. This part, which Cesalpino calls "wood", can be compared to what we now call "sapwood". It contains the basal cell layers (meristems) that produce the wood cells (xylem inside, phloem outside) and the outside bark cells (figure 11). This intermediate part is the only source of vitality for the tree. Cesalpino observes in chapter 3 (§34–37) that trees can survive even if their central part is hollow; this obliges him to develop *ad hoc* hypotheses to ensure consistency with his initial thesis, according to which vitality is derived from the central part of the stem.

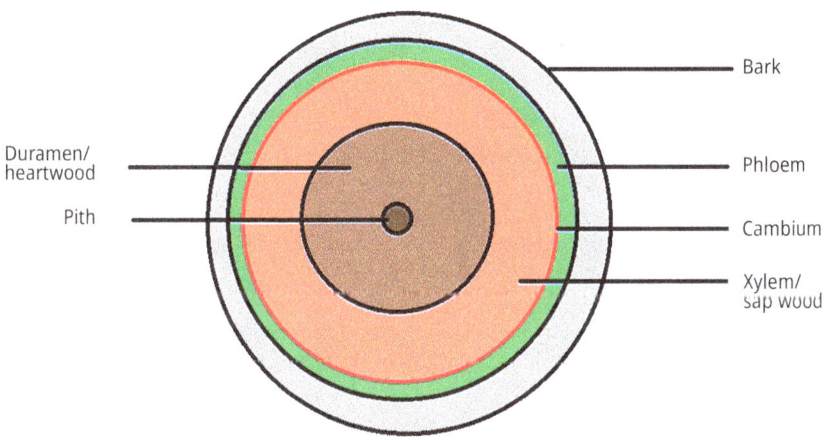

Figure 11: Current illustration of tree anatomy.

What Cesalpino calls *cerebrum* is unclear (§14). He distinguishes two types of medulla in animals: the cerebral medulla, and the spinal medulla, derived from the former. It is most likely that the brain of plants corresponds to the cerebral medulla of animals; Cesalpino locates this plant brain in the root collar, while what he calls the "pith" spreads from the neck root collar to the stems and may therefore be equivalent to the spinal medulla.

Chapter 2. Nutrition (§15–26)

Although Cesalpino's theses may appear speculative and unconvincing to a modern reader, he was, in his day, at the forefront of animal anatomy and physiology. Indeed, he combined his observation of latex flowing from euphorbias and other plants to his understanding of the blood circulatory system of animals to postulate the existence of a vascular system in plants, although their "veins" were too small to be observed with the naked eye (§17).

The conducting tissues of plants were not formally described until the 17th century, in the microscopic studies conducted by Marcello Malpighi (1628–1694), Nehemiah Grew (1641–1712), and Antoni van Leuwenhoek (1632–1723), but Alan Morton suggests that Cesalpino "must have been making attempts at dissecting plants and investigating their internal anatomy" in order to observe humors flowing in euphorbias or figs.[506] Whether or not this is the case, Cesalpino develops his own theory of plant nutrition by analogy with his knowledge of animal physiology,[507] and proposes a physiological mechanism of absorption and filtration of liquid through a principle of vital heat that animates plants, as the heart does for animals. He identifies in the plant body a functional equivalent to the animal heart, or to other organs (brain, marrow, membranes), but these analogies are not simplistic structural analogies. As discussed above, the "heart" of plants is not so much an organ whose location may be determined, as a heuristic concept deduced from its diffuse effect.

The keystone of Cesalpino's conception of life is the principle of heat, from which flow the other functions: growth, movement, and sensation. This heat may be the release of heat in the physical sense, or a metaphysical principle of life. In *Daemonum investigatio peripatetica*, Cesalpino suggests that this principle is more divine than fire:

> [...] calidum innatum corpus esse divinius quam elementa, quamvis ex illis constituatur: et merito respondere elemento stellarum non solum ratione foecunditatis, quae ab intelligentia habetur in eo comprehensa, sed etiam ratione immortalitatis subiecti, et puritatis.

> [...] innate heat is a body more divine than the elements, although it is composed of them. It rightly corresponds to the element of stars, not only as regards its fecundity, which it possesses from the intelligence comprised in it, but also as regards the immortality and the purity of its subject.[508]

However, Cesalpino explains in a circular fashion that heat is not perceptible in plants because they lack motion and sensation. The plant organism must therefore "consume" (*absumere*) less food and heat and use them for the purpose of growth alone (§16). The "proof" that the vitality of plants depends on heat lies in the observation that plants

506 Morton 1981a: 132.
507 Sachs 1875 (Engl. transl. Garnsey 1890: 450–453).
508 Cesalpino 1593: 2.148 A, on this see Blank 2023.

grow more quickly in the spring and summer sun (§24). Heat remains metaphysically necessary to account for the function of nutrition and growth in a unitary theory of life based on the animal model. Nevertheless, Cesalpino notes that our senses alone do not allow us to quantify exactly the presence or absence of heat, but only give us a relative appreciation of how cool or warm an object is. This is a first step towards the modern, scientific effort to obtain more rigorous quantitative measurements from instruments, instead of relying upon subjective appreciation by the individual, which is not generalizable and is affected by circumstances (§17).

Plant nutrition remains problematic when compared to the animal model. In the wake of Aristotle, Cesalpino states that, in animals, the selection of food depends on the sense of taste. Since he considers plants to be devoid of any sensation in principle, Cesalpino makes an unsuccessful attempt to offer another explanation for the absorption of food by plants, based on physical mechanisms such as magnetization, vacuum force, or imbibition (§20). He then resorts to a traditional elemental explanation: the nourishing soul uses the roots of the plant, which are able to absorb water because of their greater affinity with this element than with air or earth. He also draws analogies with crafts and mechanics, and refers to the combustion of oil absorbed by a wick to explain the ascent of the sap, or to capillary filtration to explain the filtration of food that occurs during the movement of ascent generated by the vital heat (§23, 24) (figure 12).[509]

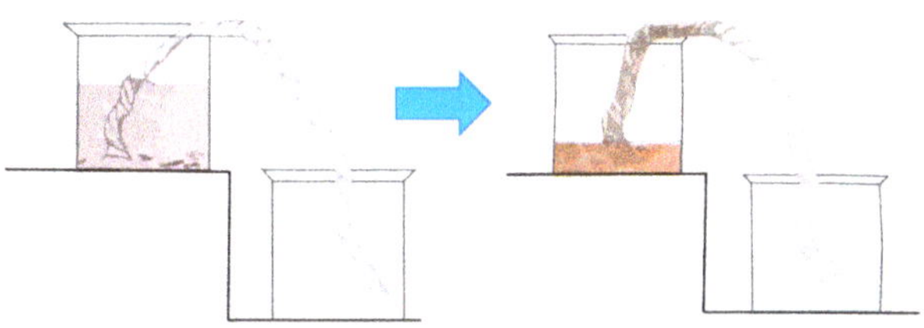

Figure 12: Illustration of capillary filtration.

Thus, Cesalpino acknowledges the important link between nutrition and plant growth, and the refinement and movement of sap, although his explanations, while probably empirically informed, are not truly experimental. Like all his predecessors, he does not differentiate between the ascending raw sap and the descending elaborated sap:[510] he does not understand the essential role of the leaves in photosynthesis and nutrition, which allows the conversion of raw sap into elaborated sap enriched with

509 This thesis of an ascending filtration of the sap in plants is repeated almost exactly in the *Metamorphosis of Plants* of Johann Wolfgang von Goethe (1749–1832) published more than two hundred years later. See in particular §27–30 of the *Metamorphosis of Plants* (Goethe 1790), and our commentary of chapter 6.
510 This distinction is due to Claude Perrault (1613–1688).

sugar, nor their role in respiration. Cesalpino's observations enable him to identify the main elements in the physiological problem of plant nutrition: sun or heat as source of energy, raw sap obtained by the roots, its movement, and its nutritional refinement. The scientific validity of his physical explanations is hindered by the lack of experimental apparatus, and he resorts to metaphysical arguments to fill the gaps unsolved by experience. As with the problem of blood circulation, Cesalpino came very close to finding the solution that he was seeking.

According to botanist Jean-Baptiste Saint-Lager (1825–1912),[511] chapter 2 of Book I of *De plantis* suggests that Cesalpino already had a conception of the circulation of blood in animals, distributed throughout the body by the arteries; according to the description he gave in his *Quaestiones peripateticae*,[512] he had observed it himself. He is credited by some for the discovery, through observation and experimentation, of blood circulation in 1593, and very nearly describes both small and large circulation in chapters 9 and 19 of the sixth book of his posthumous treatise *Praxis universae artis medicae*, published in 1606.[513] However, these passages do not clearly describe a circulatory conception of a closed blood system, nor the communication between veins and arteries. It is therefore to William Harvey that we owe the demonstration of the complete circulation of blood, returning from all parts to the heart; he successfully proposed a new explanatory logic based on experimentation.[514] And yet, Cesalpino came close to conceiving the circulatory system when he observed that the blood in the veins travelled in one direction only – towards the heart (§15) – but his use of the neologism *circulatio* in this context is misleading, because his conception is founded on the model of irrigation (suggesting that blood diffuses rather than circulates).[515] Although he emancipates himself from Galen more than any of his contemporaries, Cesalpino remains faithful to physiological ideas derived from Aristotle, and follows the qualitative approach to blood flow adopted by his contemporaries, instead of attempting a quantitative approach (like Vesalius).[516] Nevertheless, Cesalpino derived from his comparative study of plant and animal physiology both knowledge and an increased faculty to consider the fundamental questions of biology in an innovative manner. Although he may have only hinted at a theory of blood circulation rather than developing it fully,[517] the very existence of controversies surrounding the attribution of the discovery of blood circulation to him rather than to Harvey, supported by many of his observations, including passages from *De plantis*, attests the significance of the role that he played in this episode of the history of science.[518]

---

511 Saint-Lager 1885: 68–69n1.
512 Cesalpino 1571: 5.125.
513 Cesalpino 1606: 469–470, 501–503; Arcieri 1945: 93; Peller 1949; De Wit 1992: 350–353.
514 Bylebyl 1972; Capecci 1983; Wear 1990.
515 Peller 1949.
516 Peller 1949.
517 Pagel 1953; Rothschuh 1973: 67; Wear 1990.
518 Peller 1949.

Chapter 3. Development and Growth: Embryo, Bud, Shoot, and Bark (§27–37)

This chapter deals with the germination, growth, budding, and development of plants. These phenomena are grouped together by Cesalpino under the term *germinatio*, which has no suitable equivalent in English.[519] The word refers to all the processes involving vegetative plasticity, as opposed to the growth of animals, which is, in most cases, qualitatively predetermined from the end of embryogenesis. Plants, on the other hand, are characterized by what is known in today's vocabulary as somatic embryogenesis. Somatic embryogenesis is the fact that the different parts of plant bodies remain plastic and grow, not only quantitatively (like growing animals), but also through the constant renewal of organs. Cesalpino notes that, in animals, the creation of new parts is limited to secondary organs such as hair, teeth, and horns, which do not intervene directly as instruments of the soul's faculties. Therefore, he chooses to compare the development of a plant to that of the animal embryo which "grows" like a plant (§28). Hippocrates first suggested the analogy between an animal fetus and a plant in *On Generation* and *De natura pueri*, an analogy also adopted by Aristotle, Galen, and the Neoplatonist tradition.[520] It is very likely that a physician and philosopher such as Cesalpino would have read these treatises, or encountered this analogy in the works of later authors; however, he seems to be the first to refer to the animal fetus to explain the development of the plant, rather than the opposite. This reversal is significant; it initiates a shift from the central importance of medical knowledge towards the analysis of plants for themselves, and thereby confirms the originality of Cesalpino's approach, and the status of *De plantis* as one of the very first, treatises of modern botany – if not the first.

Cesalpino does not challenge the Aristotelian theory of animal generation, according to which conception depends on the external supply of male seed, but he considers that this theory is not applicable to the generation of plants. Indeed, according to Aristotle:

> Ἐν μὲν οὖν τοῖς φυτοῖς οὐ κεχώρισται τὸ θῆλυ τοῦ ἄρρενος· ἐν δὲ τοῖς ζῴοις ἐν οἷς κεχώρισται προσδεῖται τὸ θῆλυ τοῦ ἄρρενος.
>
> Now in plants the female is not separate from the male; in certain of the animals, however, it is separate, and here, in addition, it has need of the male.[521]

This thesis is justified empirically by the absence of mating in plants, and leads, by deduction, to the conclusion that there is no sexuality in plants. Indeed, botanists did

---

519 For a discussion of the concept, see Greene 1983: 2.825–831.

520 Brisson 2023.

521 Aristotle, *De generatione animalium* II, 5, 741a4–6 (ed. and Fr. transl. Louis 1961: 72; ed. and Engl. transl. Peck 1943: 201).

not recognize the existence of sexes and sexual reproduction in plants until Rudolf Jakob Camerarius (1665–1721)' experimental demonstration to the contrary in 1694,[522] and Cesalpino accordingly adopts the Aristotelian hypothesis of an internal principle of plant conception, suggesting that the seeds of plants develop by themselves. However, the Aristotelian analogy of the seed with the laid egg is misleading, for, as Cesalpino notes, although the egg seems to develop on its own, it may only do so after it has been fertilized by a male. Moreover, in the Aristotelian doctrine, the sensitive faculty in the soul originates in the male principle; but since plants are insentient, they *de facto* do not need fertilization by an external male principle. Cesalpino also compares plants to an inverted womb that produces its fetuses – the fruits – externally (§29).

The principle of "development" (*germinatio*) also applies to parts other than the fruit, such as, for example, the growth of leafy shoot buds that accompany the fruit (§ 30). Cesalpino explains that the protective role of the leaves is twofold: it provides a layer of physical protection around the fruiting flower bud, and, later, shade to the fruit, once it has formed. This idea too is borrowed from Aristotle:

> Καὶ ἐν τοῖς φυτοῖς φαίνεται τὰ συμφέροντα γιγνόμενα πρὸς τὸ τέλος, οἷον τὰ φύλλα τῆς τοῦ καρποῦ ἕνεκα σκέπης.
>
> It is evident that even in plants things come to be that further the end – for example, leaves for the sake of shading the fruit.[523]

In warm countries, where trees bear fruit all year round, the leaves do not fall off. Since the main role of the leaves is to provide protection, Cesalpino proposes that the leaves are produced by the outer bark, with which they share their nature and protective qualities. This outer bark is differentiated from the inner part of the bark, the "bast" (*liber*) (§32).

These anatomical observations of the bark are confirmed by current botany, which distinguishes the outer layer of the bark, the cork, from its inner layer, the bast.[524] However, the outer bark does not produce leaves as Cesalpino suggested; it is composed of dead cells and fulfils only a protective role.

Cesalpino also mistakenly suggests that the fruit-bearing shoot, contrary to the leaves, is not generated by the bark, but from the pith and its surroundings (i.e., the wood or the substance surrounding the central pith of non-woody plants) (§33). Current botany shows that all aerial parts of plants (leaves, stems, flowers, and fruits) are generated from meristems (groups of undifferentiated cells) present on the surface and at the end of the stems and roots. The inner wood's only function is sap conduction (and

522 Camerarius 1694. See Magnin-Gonze 2009: 111–112, 119–121.
523 Aristotle, *Physics* II, 8, 199a24–26 (ed. Ross 1950: 199).
524 The bast is still called *liber* in French.

secondary growth, i.e., widthwise in the case of trees), and pith and heartwood simply consist of dead matter.

The location of the morphogenetic power of plants – their "prolific force" (*vis prolifica*), in Cesalpino's words – can be demonstrated experimentally. If a complete ring of bark is removed from a tree or branch (girdling technique), even only a few centimeters wide, the elaborated (descending) sap ceases to feed the lower parts of the plant, and the plant withers and dies. On the contrary, removing the interior of a plant, for example, hollowing out a tree, does not prevent the plant from growing and producing new parts. Like Theophrastus,[525] Cesalpino is aware of these two phenomena, but he misinterprets them (§35, 36). He does understand that the bark plays a vital role in the survival of the rest of the plant because of its role in conducting sap, but he thinks that this vital movement operates only from the bottom upwards; indeed, the double circulatory system of the sap (raw rising, elaborated descending) was still unknown in Cesalpino's time.[526]

Cesalpino posited that the production of shoots depends on the pith and the wood. As such, he explains the vitality of hollow plants and trees (such as rushes or willows), which have no pith, by suggesting that all their vital force is concentrated in the sapwood (or in what takes its place). This idea, which grants primacy to the sapwood over the pith (or heartwood), is not unfounded. Cesalpino observes that the most central wood becomes hollow when it ceases to be irrigated by the sap provided by the sapwood veins (§35) and draws a contrast between the nature of the wood and pith, and that of the bark.

From the observation that hollow trees can produce flowers and fruit, Cesalpino develops a new explanation with reference to grafts and their outcomes. A bark graft taken from one species of tree produces the leaves, but also shoots, flowers, and fruits of that species when transferred to another species of tree (the rootstock). Cesalpino endeavors to explain why the fruiting body of the donor scion is preserved on the rootstock although, in his conception, it depends on the wood of the donor – and the scion has only bark and no wood. He explains that the graft takes because it appends itself to the underlying wood, and therefore bark grafting is able to produce fruit and not just leaves (which originate from the bark only, in Cesalpino's model). However, he also believes that the wood plays a role in fruiting, and thus an apple bark graft on a pear tree should produce pears, from the pear wood to which it is attached. Cesalpino does not go so far as to deny that an apple graft produces apples on a pear tree but supports

525 Theophrastus, *Historia plantarum* IV, 15, 1 (ed. and Fr. transl. Amigues 1988–2006: 2.114; ed. and Engl. transl. Hort 1916–1928: 1.405) and *De causis plantarum* V, 17, 1 (ed. and Fr. transl. Amigues 2012–2017: 3.51–52; ed. and Engl. transl. Einarson and Link 1978–1990: 3.181–183).

526 The upward movement of the raw sap is ensured by the xylem, the wood, while the part situated under the bark, the phloem, ensures the downward movement of the elaborated sap, loaded with sugars produced by photosynthesis after its passage in the leaves.

his theory by resorting to a convoluted pirouette: the seeds contained in the fruit borne by the graft would share the nature of the wood and pith of the rootstock (§37) from which they originate. He states in chapter 9:

> Non est autem eadem seminis materia et pericarpij; illa enim ex profundis partibus erumpit, vt superius comprobauimus: haec autem ex exterioribus magis accedit, scilicet ex cortice. [...] pericarpium materiam ex cortice sumat; nam inoculatio fructum parit pro natura corticis inoculati. Semen autem pro natura ligni, vt superius notauimus.
>
> The matter of the seed is not the same as that of the pericarp: the former comes from deeper parts, as we have already shown; the latter more from the outside, in other words from the bark. [...] the pericarp receives its matter from the bark [of the tree]; indeed, shield budding will produce the fruit from the bark that is grafted. The seed [receives its matter] from the nature of the wood, as we have remarked above (§88).

This sheds light on the more ambiguous passage in §37, in which he first introduces this thesis:

> Producuntur autem et folia et fructus secundum naturam corticis: quia haec in omnibus ex cortice ortum ducunt: semina autem interna non secundum naturam corticis, sed ligni subiecti fiunt; nam si serantur, nascuntur, non pro natura insiti corticis, sed vt plurimum syluestre genus: ortum enim ducunt ex medulla, non cortice.
>
> The leaves and the fruit are produced by the nature of the bark, because their growth is generated from the bark. On the other hand, the seeds inside [the fruit] are not made from the nature of the bark, but from the wood beneath it. Indeed, if they are sown, they appear, not as the same nature as the implanted bark, but rather like the wood: their growth comes from the pith and not from the bark. (§37).

Cesalpino seems to indicate that only the pericarp of the fruit comes from the bark, while the seeds contained in the fruit would depend on the nature of the wood that generates them. Therefore, if one were to sow the seeds of apples obtained from a graft carried by a pear tree, these seeds would produce pear trees. Experimentation easily disproves this, and indeed Cesalpino does not dwell on this thesis, which he conveys, moreover, in an elliptical manner. The following passage from Brasavola, a contemporary of Cesalpino, supports our analysis of grafting and its consequences in *De plantis*. Brasavola refers explicitly to a similar theory of hybridization by grafting, generated by combining bark and wood from different species:

> Constat enim per uariam Insitionem plura pirorum genera apud nos extare, quam ab antiquis tradantur, longus enim rerum usus, et uaria diuersarum specierum miscella uarias species inuenit et adeo creuit ars ut audeat etiam diuersorum generum insitionem tentare, quo gustus alter fructui condonetur.
>
> It is evident that through various graftings it has come to pass that more sorts of pears now exist than all those of which the Ancients wrote. The long continued practice of the art, and the blending,

> has now progressed to the point of undertaking to conjoin things belonging to different genera, and so yet other new qualities are acquired.[527]

French botanist and agronomist Henri Duhamel du Monceau (1700–1782) is generally recognized for providing the first experimental demonstration that a peach tree shield graft, containing only bark and not wood, grafted onto a plum tree, produced peach wood;[528] Cesalpino had already made this observation. Du Monceau also demonstrated that, contrary to the assumptions of Brasavola and Cesalpino, grafting does not modify the nature of plant species, nor generates hybrids: "Grafting, which is very suitable for multiplying a certain species, cannot produce, as it was believed, new species."[529] This theory of hybridization by grafting, although it seems strange to us today, must have been widely spread until the middle of the 18th century, for du Monceau to have to refute it by experimentation.

Chapter 4. Development and Growth: Heart, Stem, Soboles, and Leaves (§38–47)

This chapter focuses on stem growth and phyllotaxy. Cesalpino briefly mentions spontaneous generation "by putrefaction" (*per putredinem*), a conception that he seems to accept (§38). Aristotle developed this idea, for instance, in the *Generation of Animals*:

> Ἔστι δὲ καὶ ἐπὶ τῶν φυτῶν τὸν αὐτὸν τρόπον· τὰ μὲν γὰρ ἐκ σπέρματος γίγνεται τὰ δ᾽ ὥσπερ αὐτοματιζούσης τῆς φύσεως. Γίγνεται γὰρ ἢ τῆς γῆς σηπομένης ἢ μορίων τινῶν ἐν τοῖς φυτοῖς· ἔνια γὰρ αὐτὰ μὲν οὐ συνίσταται καθ᾽ αὐτὰ χωρίς, ἐν ἑτέροις δ᾽ ἐγγίγνεται δένδρεσιν οἷον ὁ ἰξός.
>
> The same sort of thing is found in plants too: some are formed out of seed, others as it might be by some spontaneous activity of Nature – they are formed when either the soil or certain parts in plants become putrescent, since some of them do not take shape independently on their own, but grow upon other trees, as for instance the mistletoe does.[530]

Cesalpino does not develop this theory in *De Plantis*, but refers more clearly to the animal version of the same theory in the *Quaestiones peripateticae*.[531] Theophrastus,

527 Brasavola 1536: 50, quoted and translated by Greene 1983: 2.685.
528 Plantefol 1969: 129; Lefèvre 2018: 1.284.
529 "La greffe qui est très propre à multiplier beaucoup une certaine espèce, ne peut produire, comme on l'a cru, de nouvelles espèces." (Duhamel du Monceau 1758: 2.85, our English translation; see also the article VIII entitled *La greffe ne change point les espèces des fruits* (Grafting does not change the species of fruits): 95–96.
530 Aristotle, *De generatione animalium* I, 1, 715b25–30 (ed. and Fr. transl. Louis 1961: 3; ed. and Engl. transl. Peck 1943: 9). About spontaneous generation of animals, see also Aristotle, *Historia animalium* VI, 16, 570a3–7 and 16–23 (ed. and Fr. transl. Louis 1964: 2.97, Engl. transl. Thompson 1910: 570).
531 Cesalpino 1571: V, 1, 92A–97A.

one of Cesalpino's sources of inspiration, also expounds the theory of spontaneous generation in plants,[532] although he refrains from asserting that this phenomenon is universal; he even questions the validity of the idea of spontaneous generation beyond common opinion, and calls for more studies on the subject[533].

Cesalpino then explains that during germination, the root is generated first, and the stems are connected to it via the heart, located at the root collar. In plants where the heart is divided, several stems emerge. Cesalpino try to elucidate the complex relationships between stems, heart, and root. While each heart produces a stem, and each stem depends on a single heart, he also observes that a single heart can be linked to several roots, because roots multiply independently from the heart or hearts that they produce. In this chapter, Cesalpino repeatedly asserts the primacy of the root over the other parts.

According to Cesalpino, the plant is neither an individual, nor a unitary organism; on the contrary, it can divide and continue to live. He resorts here to Aristotelian vocabulary: the plant, as an organism, is animated by a principle that is single in act, but may be multiple in potential (§40). Thus, roots can divide and produce new plants: garlic, for example, multiplies through the division of its bulb. Cesalpino describes this vegetative reproduction as reproduction "by soboles" (*per sobolem*) (§41).

The successive branching of the stems is detailed in §42. Cesalpino notes that each leaf is accompanied by an axillary bud (an eye) at the base of its petiole (at the level of the stipules), and has the power to generate new shoots, through the action of a new heart. This heart, located at a node, may itself be protected by a leaf sheath, in rushes or wheat, for example. Therefore, stem growth replicates the pattern for general growth pattern, and follows the vital principle explanation as well: it is as though a miniature plant were developing, with its own heart and stem, from the axillary bud on the main plant. This phenomenon is well known to botanists today as reiteration:[534] the repetition of the general growth pattern of the plant at the level of a bud, whose dormancy has ended, following the loss of the associated leaf or the breaking of apical dominance, for example.

Leaves grow around these new stems in patterns specific to species. Cesalpino describes three of the main phyllotaxies (§44). Firstly, the alternate phyllotaxy of reeds, with a single leaf per node. Secondly, the opposite phyllotaxy, with a pair of leaves per node, such as the decussate opposite phyllotaxy of the white horehound, whose pairs of leaves are perpendicular to each other from one node to the next (figure 13). The tips

---

532 Theophrastus, *De causis plantarum* I, 1, 1–2; I, 1, 4 (ed. and Fr. transl. Amigues 2012–2017: 1.2–4; ed. and Engl. transl. Einarson and Link 1976–1990: 1.3–7 and 9–11) and *Historia plantarum* III, 1, 6 (ed. and Fr. transl. Amigues 1988–2006: 2.4–5; ed. and Engl. transl. Hort 1916–1928: 1.191–193).

533 Theophrastus, *De causis plantarum* IV, 15, 4 (ed. and Fr. transl. Amigues 2012–2017: 2.103–104; ed. and Engl. transl. Einarson and Link 1976–1990: 2.117).

534 Hallé et al. 1978; Hallé 1999.

of these leaves, seen from above, form a square. Thirdly, the whorled arrangement of leaves, with three or more (*pluribus*) leaves per node (figure 14). Finally, in §46, Cesalpino distinguishes lateral growth originating in axillary buds from apical growth, i.e., the vertical growth that takes place at the top of the stems.

Figure 13: Plate of White horehound (*Marrubium vulgare* L.) showing a decussate opposite phyllotaxy.

Alternate Opposite Whorled

Figure 14: Illustration of the three main phyllotaxies.

Chapter 5. Vegetative Reproduction (§48–54)

Unlike reproduction by seed, vegetative reproduction involves soboles. Soboles are parts of the plant's body that, on becoming independent, become capable of creating a new plant, whereas seeds must go through the process of generation. The sobole is comparable to a fetus attached to the mother plant, while the seed is similar to the egg (§53).

As a result from his observation and analysis of vegetative reproduction, Cesalpino re-evaluates the idea that plants have a heart, a life principle – or soul, "located in a privileged place" (*in aliqua parte principali*) as it is in animals (§8). Indeed, since the prolific power of plants is found in each of their parts, the heart should be found in each of them; this explains why a cut root can produce a stem and recreate a whole plant, and vice versa. This ability of plants is still widely exploited in agriculture nowadays, in the techniques of cuttings and layering. Cesalpino notes that, in addition to this vegetative reproductive force shared by most plants, some species have parts specifically devoted to this function: the bulblets. He gives the example of the coral root, but also of the perpetual leek and of the rocambole garlic (sand leek), which produces bulblets at the top of its stems (figure 15, 16, and 17).

Figure 15: Picture of Rocambole garlic (*Allium scorodoprasum* L.) with cloves germinating at the top of the stem.

Figure 16: Plate of Rocambole garlic (*Allium scorodoprasum* L.).

Figure 17: Picture of Coral root (*Cardamine bulbifera* (L.) Crantz) showing bulblets along the stem.

Chapter 6. Seeds (§55–63)

After vegetative reproduction, Cesalpino logically addresses reproduction by seed, and describes the generation, morphology, and germination of seeds.

In 1583, there was no conception of sexual reproduction or sexes in plants, and these concepts only occurred in analogies in botanical treatises. This is consistent with Aristotle's position:

> Ὅσα δὲ μὴ πορευτικὰ [...], διὰ τὸ παραπλησίαν αὐτῶν εἶναι τὴν οὐσίαν τοῖς φυτοῖς, ὥσπερ οὐδ' ἐν ἐκείνοις οὐδ' ἐν τούτοις ἐστὶ τὸ θῆλυ καὶ τὸ ἄρρεν ἀλλ' ἤδη καθ' ὁμοιότητα καὶ κατ' ἀναλογίαν λέγεται·.
>
> The creatures which cannot move about [...] are in their essence similar to plants, and therefore, as in plants, so also in them, male and female are not found, although they are called male and female just by way of similarity and analogy.[535]

Theophrastus and Pliny sometimes referred to a tree as male or female, but usually to designate two different, but closely related, species: thus, they may describe as "male" an apparently more robust species, and as "female" a fruit-bearing species.[536] The case of dioecious plants, with separate male and female plants, is more complex, but some ancient authors had already identified a phenomenon similar to fertilization;[537] thus, Theophrastus and Pliny reported that powder (pollen) from the branches of male palms is shaken onto female trees to fertilize them.[538] Prospero Alpini (1553–1617) provided the first experimental demonstration of the fertilization of date palms in the 1580s,[539] but this was not enough to confirm the idea of sexes in plants, because these observations did not fit into a general theory of sexual reproduction. Such a theory would be proposed by Camerarius in 1594, in his *De sexu plantarum epistola*[540].

Before the 18th century, dioecious plants were considered special cases that had to be reinserted into a more general theory of reproduction that included all other plants (hermaphrodites).[541] However, the lack of any precise anatomical observation of flowers

---

535 Aristotle, *De generatione animalium* I, 1, 715b16 and 18–21 (ed. and Fr. transl. Louis 1961: 3; ed. and Engl. transl. Peck 1943: 7–9).

536 Saint-Lager 1884: 8–20; Negbi 1995: 317–332; Bretin-Chabrol and Leduc 2009: 205–223.

537 Saint-Lager 1883–1884: 1–8. He mentions Herodotus as one of the oldest sources on the reproduction of date palms: Herodotus, *Histories* 1, 193 (ed. and Engl. transl. Rawlinson 1858–1860: 86).

538 Theophrastus, *Historia plantarum* II, 8 (ed. and Fr. transl. Amigues 1988–2006: 1.64–66; ed. and Engl. transl. Hort 1916–1928: 1.151–153); *De Causis Plantarum* II, 9 (ed. and Fr. transl. Amigues 2012–2017: 1.89; ed. and Engl. transl. Einarson and Link 1976–1990: 1.279–281); Pliny, *Naturalis historia* XIII, 7 (§35) (ed. and Fr. transl. André et al. 1947–2015: 13.29; ed. and Engl. transl. Andrews et al. 1938–1962: 4.119).

539 Alpini 1592: 4.ch. VII; Premuda 1987: 71–81.

540 Camerarius 1594.

541 Delaporte 2011: 30–31.

prevented the development of a scientific theory of plant reproduction. Renaissance botanists inherited from Antiquity the theory according to which the seeds that fall into the soil are sterile until they are fertilized by a "breath" (*spiritus*), manifested as a wind called Favonius or *spiritus genitalis*. Ruel presents this theory in *De natura stirpium*:[542]

> Primus annuo ordine naturae conceptus est, postquam incipit perflare Fauonius: exinde Februario. hoc enim maritantur uirescentia e terra, hic est genitalis spiritus mundi [...]
>
> The first [plant] of the year in the order of Nature is the one conceived after Favonius begins to blow, that is after February. For then, they are impregnated by the verdant earth, that is the fertile spirit of the world [...][543]

Favonius is the name of the Roman deity of the west wind, equivalent to the Greek Zephyr, a warm wind associated with spring. Pliny explains that Favonius fertilizes what lives in the earth by causing the germination of seeds and brings life to plants.[544] Ruel is faithful to Pliny in his description of Favonius as the breath of the world: it blows all year round, but with effects that vary according to place and season.[545] One of these effects is the budding of male and female plants:

> Primo fauonio germinat cornus, tam mas quam foemina [...]
>
> With the first favonius, the cornel grows, the male as well as the female [...][546]

For Ruel, male cornel and female cornel were likely two different species, and his explanation of plant fertilization and germination therefore included no mention of, or even allusion to, a sexual process.

Cesalpino's interpretation is more subtle. He suggests that a fertilizing breath would also be at work between the male and female plants of dioecious species (§73).[547] According to his analysis, an excessive consumption of heat explains why the male flower of dioecious species does not produce fruit, but the male emits a breath that enables the flower of the nearby heat-deficient female plant to bear fruit.

In the 16th century of Ruel and Cesalpino, plant reproduction was still studied from a metaphysical perspective, and the role of flowers was not yet understood.[548] Dioecious

542 Ruel 1536: 1.28–33; see also Greene 1983: 2.648–652.
543 Ruel 1536: 1.28.
544 Pliny, *Historia naturalis* XVI, 39 (§93) (ed. and Fr. transl. André et al. 1947–2015: 16.50; ed. and Engl. transl. Andrews et al. 1938–1962: 4.449).
545 Ruel 1536: 1.28.
546 Ruel 1536: 1.29.
547 This theory of a fertilizing wind was also used to explain the reproduction of birds: see Zirkle 1936.
548 The early modern theories of plant sexuality based on empirical observations only began to develop in the second half of the 17th century. See Delaporte 2011: 101–148.

plant reproduction was analyzed as a particular case implying *ad hoc* hypotheses. Indeed, Ruel considered that there was no interaction between male and female flowers of dioecious species, notably fig trees; male and female plants were independently affected by Favonius. Cesalpino demonstrated a deeper understanding of this topic, and surmised that some interaction between male and female plant was required, operated through the exchange of heat. In this sense, dioecious plants are at the origin of their own fecundation, and the process of reproduction is not ascribed to the external power of a transcendent cause, even though this heat seems more metaphysical than empirical in Cesalpino's physiology.

Conflicting theories of plant reproductions arose in the 17th and 18th centuries. Some attributed the fertilizing power to the soil, others considered that the plant fertilized itself. Botanists such as Nehemiah Grew and John Ray defended the existence of male organs to fertilize the female organs, and inspired Camerarius to envisage the empirical sexual reproduction of plants, in which fertilization was operated by the transfer of pollen from the stamens to the pistils of the flowers.[549]

In the context of the Renaissance, it is probably an exaggeration to see Cesalpino as an advocate of sexual reproduction or even of the existence of true male and female sexes in flowers, as it has sometimes been argued[550]. In fact, except in the case of the dioecious plants, Cesalpino explicitly asserts the absence of sexes in plants. He also rejects the intervention of a male principle to explain the viability of seeds: only (female) matter is considered necessary for their development (§57). He even adds that the terms "male" and "female" are only used to describe (dioecious) plants by analogy with the animal world. In chapter 3 (§37), he seemed to defend the thesis of hybridization by grafting, which allows him to explain the great diversity of cultivated species and varieties without resorting to the idea of crossing by sexual reproduction. Cesalpino contributes nonetheless to the history of the discovery of plant sexuality as a precursor of flower morphology. He describes the different parts of flowers in chapter 7; the differentiation between pistils and stamens and their precise description constitutes the first step towards the identification of their respective functions.[551]

As mentioned above, Cesalpino considers the seed as equivalent to an already fertilized egg. Indeed, dissection of a seed reveals that it contains a germ, which is "like a little heart" (*quasi corculum quoddam*) and resembles an eye, and a white substance which sustains the germ. Cesalpino calls this substance the "pulp" (*pulpa*) – it is in fact

549 Delaporte 2011: 101–148; Daugey 2015: 81–108.

550 As did, for example, famous Swiss botanist Augustin-Pyramus de Candolle (1778–1841), one of the founders of plant biogeography (Candolle 1832: 48).

551 According to the French botanist Raymond Dughi (1932–), who worked as a lichenologist in the Muséum d'Histoire Naturelle d'Aix-en-Provence, Linnaeus' sexual system is deduced from Cesalpino's views on flowers and on the priority of reproductive organs to elaborate a classification. Linnaeus also shared Cesalpino's Aristotelianism (Dughi 1957: 164; 181).

the kernel, known today, in a more technical sense, as the endosperm. These two parts are essential and interdependent (§58).

According to Cesalpino, seeds are generated by refined food and are an extension of the pith (or of the part of the plant between the bark and the pith, the sapwood, when the pith is missing). The purest part of the food that reaches the fruit creates the pith of the seed, and the rest generates the bark (§56).[552] The consistency and nature of the seed are explained through the theory of the four elements, i.e., through its affinities with water, heat, and air (§60). Although this reference to elemental principles may seem metaphysical, it is supported by empirical findings, such as the observation that seeds must dry out to reach maturity, and that this dehydration makes them likely to absorb water from the soil in order to germinate (§61). Similarly, the vital heat principle necessary for the seed to germinate in the soil is aroused by the reaction of water and seed, which would give off heat like the reaction of water and quicklime (§61).

In §62, Cesalpino accurately describes the stages of germination of a seed (a dicotyledon): growth of the radicle, then erection of the hypocotyl and unfolding of the cotyledons, which differ in nature from the subsequent leaves.[553] He then mentions another type of germination, which is difficult to identify precisely, and allows two interpretations. Cesalpino may be establishing a distinction between the normal germination of the seed – mentioned above – and an aborted, failed germination, leading to the desiccation of the cotyledons; or he may be providing a technical description of the germination of a monocotyledonous seed in relation to the germination of a dicotyledonous seed – analyzed at the opening of the section. In monocotyledons, the cotyledons do not emerge from the ground, but remain in the seed and serve as internal food for the shoot, directly generating a true leaf from the seed (figure 18). Unfortunately, the text is too ambiguous to decide in favor of one interpretation over the other. This distinction between mono- and dicotyledons is not found explicitly in Cesalpino's works, and is not used in his classification,[554] although it was fundamental to the classification proposed by John Ray in 1682.[555]

552 According to Dughi (1957: 163), famous 17th-century anatomist Marcello Malpighi developed the theory of sap purification – the purest part goes to the seeds. Tournefort, better known for his plant classification system based on corollas, endorsed Malpighi's theory. Philosopher and amateur botanist Goethe developed a similar idea in his *Metamorphosis of plants*, stating that flowers are produced thanks to a more refined sap (Goethe 1790: §27–30). However, *De plantis* indicates that Malpighi did not initiate these ideas.

553 For a commentary on germination in *De plantis*, see Bremekamp 1952.

554 Sachs 1875 (Engl. transl. Garnsey 1890: 57–58). Morton 1981a: 134–135, and Magnin-Gonze 2009: 77 argue that the distinction between mono- and dicotyledonous plants exists in *De plantis*, since Cesalpino contrasts bifid seeds (*semina bifida*) with others (§62). In our opinion, the text does not allow for such an assertive interpretation.

555 Stevenson 1947: 254; Raven 1950: 193–194. On Ray, see also Magnin-Gonze 2009: 107–111, 123–124.

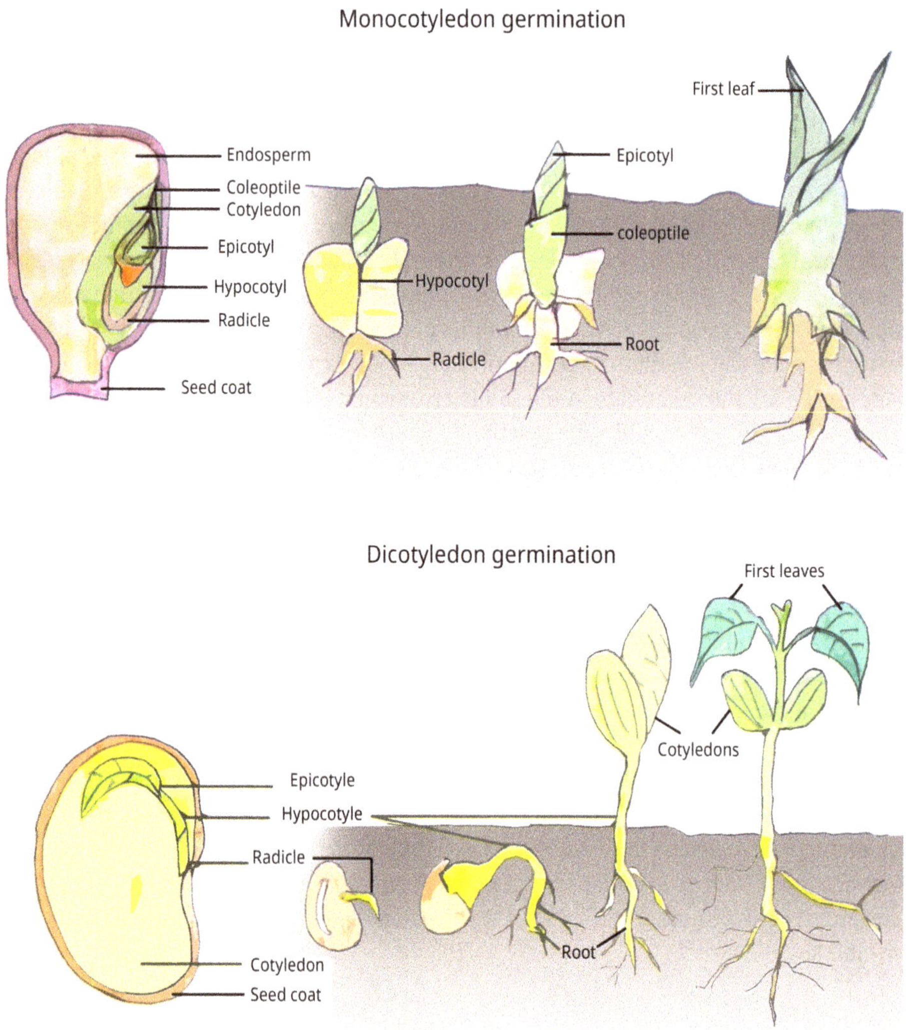

Figure 18: Comparison between monocotyledon and dicotyledon germination.

Chapter 7. Flowers (§64–76)

Cesalpino considers that the function of flowers is to protect the seeds in their early state, just as the leaves of the shoots provide shade for the fruit (§67), and that flowers originate from the same vital breath that generates the seeds. The subtle and light substance of this breath explains the fragrant exhalations of the flowers (§65). In addition to the fragrance, the sweet nectar present in the flowers (and the honey that is

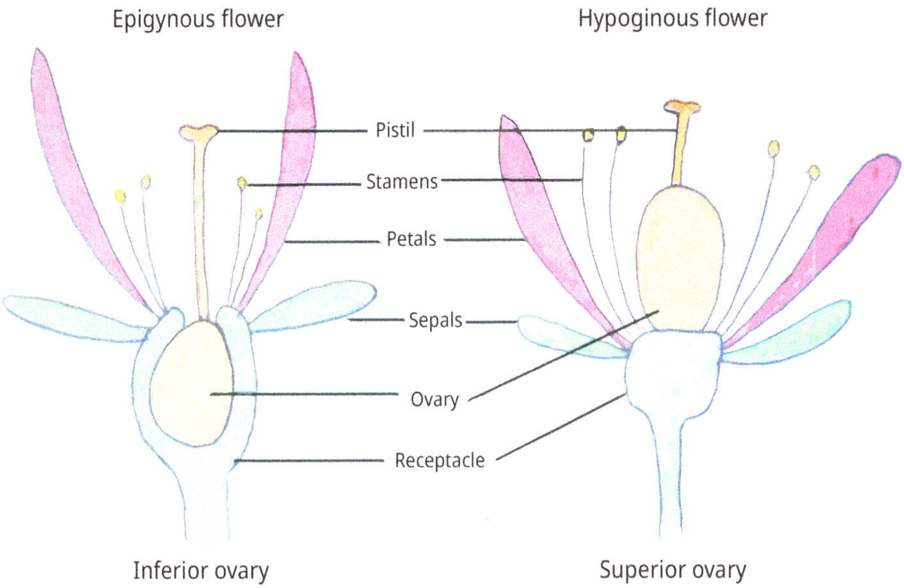

Figure 19: Types of flower and ovary implantations.

produced from it) is another clue to the nature of the substance from which flowers and seeds originate (§66). Cesalpino considers the seeds to be the *causes* of the flowers, rather than the opposite; this reveals that he does not understand the true function of flowers (§67), and indeed he states that the flowers are "fed [...] by the nourishing [substance] that comes together with the seed" (*nutriantur [...] alituosa, quae una cum semine erumpit*) (§67).

In §67, Cesalpino may attempt a distinction between species of epigynous or perigynous flowers (with an inferior ovary), and species of hypogynous flowers (with a superior ovary) (figure 19); unfortunately, he selects as examples species that do not correspond to this botanical structure. Among the species mentioned – olive, plum, rose, apple, pear, and almond – the olive is the only plant with an inferior ovary. Several hypothesis can be suggested to solve this difficulty. The original text may be unclear, and Cesalpino is presenting a distinction that does not correspond to a current botanical structure. If Cesalpino is indeed describing the epigynous/hypogynous distinction, he may have chosen inappropriate examples to illustrate it, or our identification of the species that he mentions may be erroneous – this is unlikely, because the species in question are very common.

In the rest of the chapter, the Italian botanist indicates that flowers have leaves (*folia*), but that these are different from the leaves of the shoots; indeed, the word *folia* here refers to the petals of the flower's corollas. The leaves of the stems remain in place and protect the fruit as it matures, whereas the petals (the leaves of the flowers) protect the fruit only as it forms, which is why they fall off as soon as it starts to grow.

There was no technical vocabulary for floral morphology in Cesalpino's time, and the current term "petal" is absent from the vocabulary of *De plantis*, although Cesalpino clearly identifies the organ. According to the Centre National de Ressources Textuelles et Lexicales (CNRTL), French botanist Antoine de Jussieu (1686–1758)[556] first defined the term "petal". It is borrowed from the botanical Latin *petalum*, first noted in a work by Fabio Colonna, and itself derived from the Greek πέταλον, "leaf of a plant, flower, tree".[557] Although Colonna's concept of "petal" (*petalum*) may have been more technical, as evidenced by the great precision and quality of his illustrations, the term only appears in a posthumous edition, posterior to the publication of *De plantis*. Moreover, the term "petal" was not truly included in botanical vocabulary until John Ray clarified its definition.[558]

The use of the term *folia* to designate the petals of the corolla does not necessarily denote a vague conception of the morphology of flowering plants. It probably did not escape Cesalpino that the morphological structure of the petals, and especially the sepals, was very similar to that of real leaves. Goethe and de Candolle later developed the theory of the metamorphosis of the various floral organs from the leaf structure;[559] this theory is not a notable innovation when one considers Cesalpino's contribution.

Indeed, the evolutionary innovation represented by flowers (petals, stamens, pistils, etc.) can be explained by a gradual transformation of leaf organs from older plants to Angiosperms. Cesalpino was not a transformist before his time, and therefore was not aware of the mechanisms underlying this transformation, and yet he observed the structural proximity of leaves and petals – similarly, zoologist Edward Tyson (1651–1708) was able to observe the similarity between the skeletons of a chimpanzee, an orang-outan, and a man, but could not explain it from an evolutionary perspective.[560] The ambiguity between a leaf and a petal in *De plantis* therefore lies in terminology rather than in morphology, since Cesalpino lacks above all a technical vocabulary. The technical definition of botanical terms only progressed under the impetus of Charles de l'Écluse (Clusius), whose work is, for the most part, slightly posterior to Cesalpino's.[561]

And yet, while Cesalpino notes the morphological proximity of leaves and petals, his physiological explanations are strange. He considers that the leaves are generated

556 Antoine de Jussieu 1718: 18.

557 CNRTL: "pétale". The Centre national de ressources textuelles et lexicales (CNRTL) refers to the Richelet dictionnary (1759: 3.117) for the origin of the word. Richelet states that Colonna used the word for the first time in a posthumous book printed in Rome in 1649. However, Colonna had already used the Greek word in his *Phytobasanos* (Colonna 1592: 1).

558 Ray 1693: 1.XIX; Stearn 1966: 32. Ray declared to use the word petal following Colonna (Raven 1950: 194).

559 Arber 1950: 33–58; Greene 1983: 2.820.

560 Tyson 1699. Tyson is considered as the founder of animal comparative anatomy.

561 Charles de l'Écluse is famous for his accurate descriptions. He published his master work, *Rariorum plantarum historia*, a compilation of his other writings, in 1601. On this botanist, see Arber 1912: 74–78; Egmond et al. 2007 and 2010; Ogilvie 2006: 41–47, 184–191, 248–254; Magnin-Gonze 2009: 69–73.

and nourished from a liquid that acquires its green color from the sun (the sap), and he explains the variety of the colors of flowers by suggesting that the flowers are generated by, and composed from, a breath that is more sublime and powerful than the green liquid of the leaves:

> Hoc enim est spirituum sublimatorum opus [...] eorum [florum] materia non humor est, sed potius spiritus.
>
> It is in fact the work of sublime breath [...] its [i.e., flower's] substance is not a liquid, but a more powerful breath (§69).

Cesalpino states that the colors and patterns of the flowers are created by these exhalations, from which the colors and mineral veins of the subsoil also originate. He thus entertains two explanations for the colors of plants: a physiological explanation related to the sap for the leaves, and a metaphysically oriented explanation related to the breath for the flowers.

In the next section (§70), Cesalpino offers a morphological analysis of the flower supported by innovative observations. In addition to the petals, he identifies "filaments" (*stamen*) and "flakes" (*floccus*) in the flower, centered and implanted on the "seat", or "implantation", (*sedes*) of the seeds (i.e., the ovary). He also observes and exemplifies the difference between polypetal and gamopetal corollas. The former is composed of several separate petals, while the latter is composed of a single petal – the evolutionary result of the fusion of the base of several petals (figure 20). In this chapter, Cesalpino draws inspiration from Theophrastus, who had already described and differentiated gamopetal and polypetal flowers, as well as hypogynous and epigynous flowers.[562]

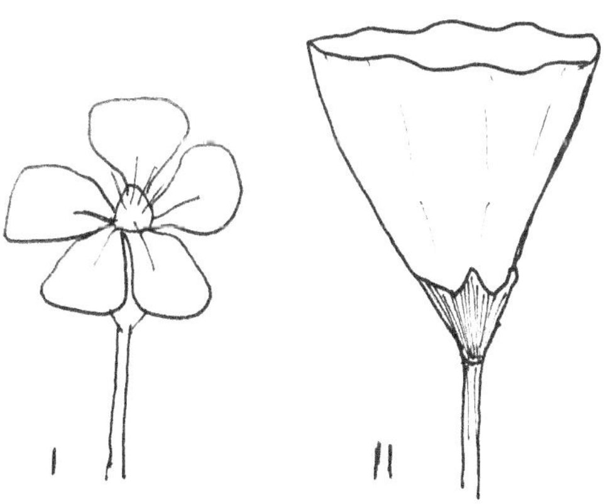

Figure 20: Drawing of Polypetal and gamopetal corolla.

562 Theophrastus, *Historia plantarum* I, 13 (ed. and Fr. transl. Amigues 1988–2006: 1.37–40; ed. and Engl. transl. Hort 1916–1928: 1.89–97).

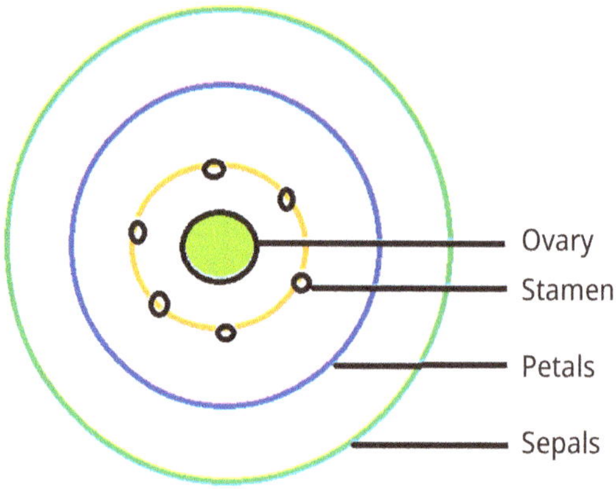

Figure 21: Diagram of floral rings.

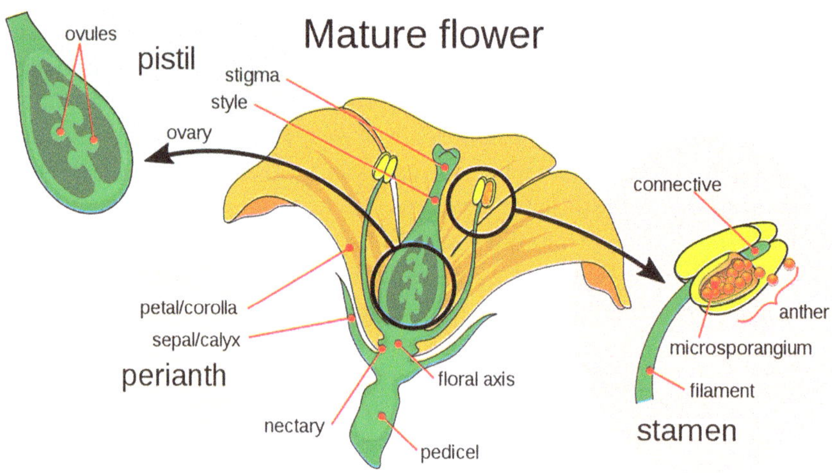

Figure 22: Diagram of a flower. Credits: Mariana Ruiz LadyofHats, 2007.

Greene notes that Cesalpino distinguishes three of the four floral circles identified today (figure 21 and 22). The first circle is the corolla, which Cesalpino describes as consisting of the colored petals. The two inner circles comprise the stamens and the carpels (pistils). However, like his predecessors, Cesalpino does not recognize the calyx and its sepals as floral parts because of their resemblance to leaves. He considers the calyx not as the fourth ring of floral organs, but as a "protection of the flower" (*tegmen floris*) (§98). It protects the fruit and is of a different nature from the flower.

According to Cesalpino, the bark is the physiological origin of the calyx and of the leaves (§74), while the flower itself originates from the layer between the bark and the pith, just like the fruit and the seeds. This hypothesis does not prevent Cesalpino from

Figure 23: Picture of a Green Hellebore (*Helleborus viridis* L.).

describing several types of calyxes and their role in protecting the fruit (§74, 75). He also identifies borderline cases: flowers with green petals, such as the green hellebore (figure 23), or plants with tepals, i.e., where the petals and sepals are identical, such as the ornithogalum, which has white tepals on the top and green on the bottom (figure 24) (§75).[563] If his physiological theory assumed that these organs resulted from the same production process, he may have concluded that both sepals and petals were transformed leaves.

Cesalpino also seems to have distinguished between the stamens and the pistils of flowers. Saint-Lager[564] considers that what Cesalpino calls the *flocci* are the stamens in general (i.e., the male organs), and uses *stamen* (*stamina*) to refer to the pistil (i.e., the female organ) in the center of the corolla, above the ovary. Greene shares this interpretation,[565] as does Morton, who also underlines the lack of coherence in the use of the two

563 Of course, the word "tepal" is not used by Cesalpino any more than the word "petal".

564 Saint-Lager 1884.

565 Greene (1983: 2.819) states that Cesalpino's *stamen* "unmistakably" refers to the pistil.

Figure 24: Picture of ornithogalum (*Ornithogalum umbellatum* L.).

terms.[566] Indeed, Saint-Lager, notes that the meaning of the word “stamen” was not yet fixed at the beginning of the 16th century.[567] Greene remarks that Cesalpino focused his systematist’s attention on the anatomy of fruits and seeds, and on the corolla of flowers, rather than on the other leaf parts, thus anticipating the approach of Joseph Pitton de Tournefort[568] – an approach derived from Cesalpino’s, according to Greene.[569] Ruel was the first clearly to distinguish stamens from pistils, even though he did not know their function;[570] however, Greene points out that what Ruel called *stamen* is in fact the pistil, and that he reserved the term *apices* for the stamens.[571] Although Cesalpino may have been influenced by Ruel’s work, which he knew, Greene may have projected his own

566 Morton 1981a: 161.

567 Saint-Lager 1884: 6. See also Sachs 1875 (Engl. transl. Garnsey 1890: 379–380); Plantefol and Prévost 1962: 156. According to Saint-Lager, the term ‘anther’ was first used by Manardi in his *Epistolae medicinales*, but did not percolate into the later botanical vocabulary. Saint-Lager probably takes this erroneous idea from the prussian physician, botanist and historian of botany Kurt Sprengel (1766–1833) (1807–1808: 1.310). Greene (1983: 2.594–595) explains, with quotations, that the term *anthera* was used before, but that it referred to a remedy for infections of the mouth, which Manardi explicitly writes (Manardi 1529: 37–38). The confusion arises from the fact that the main element of the elixir in question was taken from the stamen of the rose.

568 Tournefort 1694 and 1700–1703. On Tournefort, see Greene 1983: 2.938–966; Heim 1957; Magnin-Gonze 2009: 120–121, 128–133.

569 Greene 1983: 2.821–822.

570 Ruel 1543: 1.5. See Greene 1983: 2.627–628, 994.

571 Greene 1983: 2.628.

reading of Ruel onto Cesalpino by supplementing the text of *De plantis* as it does not mention Ruel's *apices*. Indeed, the text of *De plantis* remains uncertain; §70 states:

> Constant autem flores plerique ex folio, et stamine et floccis.
>
> Most flowers are made up of leaves [petals], a filament and flakes.

The use of the singular *stamine* rather than the plural, and thus the mention of a single *stamen* rather than several, encourages the interpretation of the term as referring to the pistil, since most flowers do have a single pistil and several stamens. However, the plural is used in a subsequent sentence of the same section:

> Stamina in medio tantum foliorum sunt ex summa parte fructus egredientia.
>
> The filaments are so well centered within the leaves [petals] that they rise above the greater part of the fruit.

Stamina here seems to refer to the multiple stamens. As such, our analysis of the text suggests that Cesalpino did not have a clear conception of what is currently identified as the pistil and the stamens.

As for the *floccus* (literally, the "flake") of the flower, we submit that it refers to the anther of the stamen, and not simply to the stamen in general, as the above commentators suggested. Indeed, the "flake" seems to designate the flaky bulge that contains the pollen. Cesalpino compares it to the sooty mushroom that forms at the end of a wick (figure 25).

Figure 25: Illustration of a sooty mushroom at the end of a wick candle.

The interpretation according to which Cesalpino identified distinct male and female organs (stamens and pistils) is not self-evident and may even constitute a historicist reading. Indeed, since Cesalpino does not recognize sexuality in plants (see chapter 6), he may also not identify two truly distinct types of organs in the middle of the flower, even if he acknowledges a morphological difference (as Ruel did). He might confuse, or at any rate equate, stamens and pistils, or, more precisely, the "filament" of the stamen

Figure 26: Plate of Meadow-rue (*Thalictrum aquilegiifolium* L.).

and the "style" of the pistil, under the term *stamen*. He also attributes a breathing or exhalation function to the flakes (§70), which shows that he had no conception of the sexual role of flowers. We have therefore opted for a translation of *stamen* as the "filament" (of the stamen, which does not formally exclude the style of the pistil) and of *floccus* as the "anther" of the stamen. Such a translation fits the rest of the work, and Cesalpino's description of the *flocci*:

> Flocci [...] pendent autem corpuscula quaedam ex tenuissimis filamentis [...] in quibus semina numerosa sunt.
>
> The flakes [...] are attached like small bodies to very fine filaments [and] containing many "seeds" (§70).

These seeds can be interpreted as pollen grains. The translation of *flocci* as "stamen" suggested by Saint-Lager and Greene is meaningless here, since stamens (filaments and anthers) cannot be attached to very fine filaments, whereas observation of the organ indeed shows that the anthers are attached to filaments.

Cesalpino also differentiates corolla flowers from "mossy flowers" (*muscosos*

flores) (§71), which are mainly composed of stamens, such as in the meadow-rue (*Thalictrum aquilegiifolium* L.) (§70) (figure 26). However, he does not consider catkins as true flowers, but as organs that replace flowers in amentifers. Although he compares the two types of inflorescences and their respective parts, it is unclear whether he views them as functionally equivalent (the long part of the catkin plays the role of the stamen) or simply as spatially equivalent (the long part of the catkin occupies the place of the stamen) (§71). Cesalpino does not discuss the female inflorescences of certain dioecious plants (nettle, mercury, hemp); this suggests that his conception of floral morphology may be limited to petalled flowers (§73).

We must therefore agree with Greene that Cesalpino's floral morphology and vocabulary are not always clear:

> It was not well digested in his own mind [...]. In formally indicating three out of the four floral circles he had advanced a stage beyond any of his predecessors, yet he was unable to apply well and consistently his own definitions.[572]

Chapter 8. Fruit and Seed Coats (§77–85)

This chapter deals with the anatomy of fruits, seeds, and their protections. Cesalpino offers a botanical definition of fruit as

> ... quod ex semine et semen continentibus corporibus constat.

> ... that which is composed of the seed and the bodies which contain it (§77).

This includes the seed, the pericarp and the bark/rind or skin of the fruit. The fruit is therefore not limited to its edible part. Moreover, in some "fruits", the part that is eaten is the pericarp, whereas in others, such as dried fruits, it is the seed. Here again, Cesalpino sets aside popular usage and proposes a modern technical definition.

In the anatomy of the seed, he distinguishes between the inner part, which houses the vitality of the seed, and the outer part, the bark, which protects it. The appearance of the seed's bark generally depends on the appearance of the seed itself, but some bark is irregular, such as that of the goat's-head (figure 27) (§83).

In §61, Cesalpino rightly pointed out that the seeds undergo desiccation under the influence of the sun's heat, which enhanced their ability to absorb liquid from the soil in order to germinate. He reasserts this process in chapter 9 (§86), in which he explains that the maturation of the seed occurs through the reaction of liquid with the heat contained in the seed (the process of coction), and that the seed then stabilizes by drying out.

572 Greene 1983: 2.824.

Figure 27: Picture of a Goat's-head (*Tribulus terrestris* L.)'s seed.

So far, Cesalpino's physiological intuitions are scientifically founded, since seeds normally go through a process of desiccation that allows them to be preserved, sometimes for several years. Germination therefore requires rehydration of the desiccated seed, not conservation of its moisture.

Yet, in §78, Cesalpino writes that the vitality of the seed responsible for germination depends on "a fertile liquid" (*humor quidam foecundus*) that must not evaporate or escape through a wound (§78). This idea of the conservation of the seed's moisture is apparently in contradiction with the statements from §61 and §86. This contradiction may result from Cesalpino's simultaneous adoption of two explanatory registers: while induction from empirical data leads him to observe the desiccation of seeds, the influence of the theory of elemental affinities may induce him to postulate the necessity of the permanence of a fertile liquid for metaphysical reasons. The contradiction may also be resolved by assigning the two statements to two distinct stages of the same process: maturation from a fertile liquid is an essential first stage for the seed; desiccation occurs later. As such, the seed's moisture may not evaporate *before* or during maturation, at the risk of abortion. However, this is not explicitly stated in the text.

Cesalpino repeats the comparison of seeds to animal eggs and identifies three different textures of the bark: membranous (thin – a grain of wheat, for example), cartilaginous (soft like the bark of a chestnut), and bony (hard like the shell of a walnut or the pit of a peach) (§79). The bark may contain several seeds, but each seed is additionally contained in a thinner inner membrane (§80).

In §81, Cesalpino explains that the seed germ, which contains its heart, is connected to the bark by a kind of umbilical cord that brings it food. He may be referring to the funiculus of the seed, i.e., the duct that connects the seed to the inner wall of the fruit at the level of the petiole (figure 28); he assimilates this inner wall to the bark of the fruit. He states (§82) that the heart of some seeds, such as the chestnut, protrudes from the bark. This heart of the seed is probably the place from which the germ emerges. In Cesalpino's conception, the seed is formed and nourished from the pith of the stems; since the pith is the innermost part of the stems, this would explain why only the

thinner scions of the year usually bear fruit. Cesalpino suggests that the veins that carry the food from the pith to the fruit are not able to traverse thick layers of wood, and thus cannot come directly from the pith of the trunk or from large branches, except for a few species.[573]

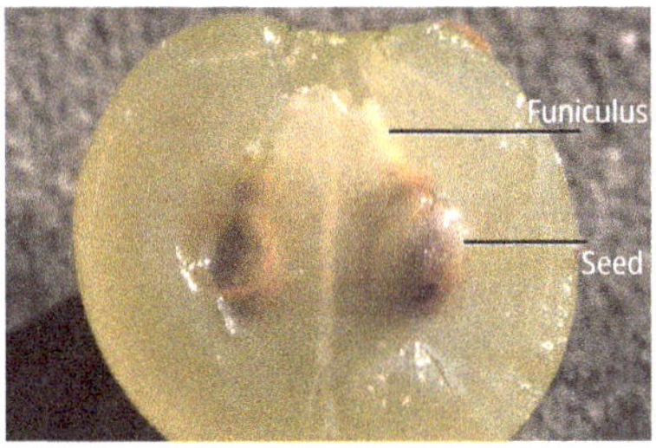

Figure 28: Section of a grape showing the seed and its funiculus.

Cesalpino's analysis of the color of seed bark in §84 has wide implications for his physiological conception. He believes, for example, that the dark color of the bark indicates its excremental nature, in contrast to the white kernel that it contains. This implies that the seed itself is created out of the purest part of the food received from the pith, while the bark is formed out of the waste, or less pure part of the food (the food coming from the seed bark could in fact be derived from wood; see our commentary on chapter 9). The intermediate-quality food from the pith is used to feed the flowers, and explains their color, between the purest white of the seed's interior, and the most impure black of its bark. Cesalpino adds that the white color of the seeds, flowers, or leaves results from the lack of contact with the ambient air; this is also the case for the pericarp (inner side), for example in the pomegranate (§89). Thus, the buried leaves of endives, or those leaves inside buds, are white because they have not been exposed to the ambient air.

This explanation is empirically unconvincing: it does not account for the white color of flowers that are exposed to the air, nor for the role of light in the growth of plants, nor for the part played by chlorophyll in the acquisition of (green) color. The central role of light in the normal development of plants was only formally established in 1780 by Jan Ingenhousz, (1730–1799), who deduced photosynthesis from the work of Joseph Priestley (1733–1804). Cesalpino had observed the influence of sunlight on the acquisition of fruit color (§89), but he did not link this observation to the acquisition of color in leaves or other parts.

573 These exceptions are called cauliflorous trees in current terminology. Cesalpino also mentions the phenomenon in §111 – see our commentary of chapter 10.

Chapter 9. The Pericarp (§86–92)

According to Cesalpino, the husk preserves the heat and internal humidity necessary for the seed to mature through the process of coction, and the pulpy pericarp of fruits containing seeds is comparable to poultices used to "ripen" tumors (§86). Cesalpino then returns to his physiological theory of interlocking layers, according to which the pericarp is derived from the bark, like the leaves and shoots. To support this idea, he refers to grafting, and more specifically to shield budding, which produces fruits of the same species as the grafted bark. He also observes that the pericarps, before they ripen and acquire their colors, are green like the leaves also produced from the bark. The seeds, on the other hand, are derived from the inner pith, just like the flowers. Cesalpino does not provide a full explanation for the origin of the seed bark (the tegument, spermoderma, or, more familiarly, the shell or seed core). He seems to consider that the three successive layers of the fruit – the seed, its bark, and the pericarp – correspond to the three successive layers of the stems: the pith, the wood (sapwood), and the bark (§88). This should imply that the bark of the seeds is derived from the wood, but Cesalpino does not assert this. The ambiguity arises partly from the implicit nature of this proposition, but also from the lack of a clear explanation of the relationship between what Cesalpino calls pith and wood. These are sometimes presented as two layers of distinct natures, sometimes as two aspects of the same layer – in §35, this was advanced as the reason why hollow trees, devoid of pith, were able to bear fruit: because their vitality depended above all on their wood.

Cesalpino then reviews the types of pericarps according to the aqueous, honey-like, or oily liquid that they contain:

> Inest quibusdam succus aquosus, qui facile putrescit, vt Cerasiis, Persicis, et tandem fructibus aestiuis, quos Orarios vocant: Quibusdam melleus et hi exsiccati diutius conseruantur, vt Ficus, Vuae. Quibusdam pinguis, ex quibus oleum extrahitur, vt ex oliuis.
>
> Some contain a watery juice, which rots easily, like cherries, peaches and all those summer fruits that are referred to as 'coastal'. Some [contain a juice that is] honey-like: once they have been dried for a whole day, they keep well, like figs and grapes. Some [contain a juice that is] oily, from which oil is indeed extracted, such as olives (§90).

He also notes that some seeds have an individual pericarp, such as pomegranate seeds, while other seeds, such as melons, have a common pericarp. Finally, citrus fruits present an intermediate case, since their segments contain several seeds and are contained in the same fruit (§90). Cesalpino adds that pericarps need water to develop, although some, such as the husks of walnuts (i.e., the epicarp and mesocarp from which walnut stain is extracted), detach themselves from their seeds (the nut itself), or rot once the seeds have matured, as in the case of squash. Thus, pericarp rot is not a problem provided the seeds have matured beforehand.

Chapter 10. The parts dedicated to the protection of the fruit and fructification (§93–113)

Some plants, especially among herbaceous species, and all conifers, do not have a fleshy pericarp. Instead, conifers have a drier envelope that plays the same role (§93). Cesalpino also mentions grains such as spelt or barley, which are wrapped in a skin that is difficult to separate.

This skin, probably the glumes and lemma typical of cereals, surrounds the florets of the ear, and later, the ripened grain (figure 29). In today's terminology, the grain is said to be naked when its pericarp, consisting of the glumes and lemma, is detached after threshing. This is not the case for so-called hulled grains (which must undergo an additional hulling process). Cesalpino's terminology seems to be different, however, since he describes as "bare" (*nudum*) grain from which only the husks are removed (§95).

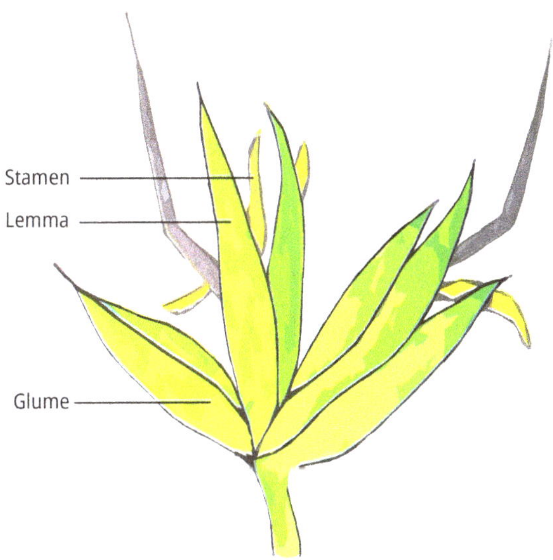

Figure 29: Diagram of a wheat spikelet.

This chapter is also an opportunity for Cesalpino to repeat that these husks protect the seeds, just as the flowers or leaves do. He calls the calyx, which sometimes remains around the seeds, "floral protection" (*tegmen floris*) (§98); it may be more or less adherent to the ripe fruit (§103). The same flower, and therefore the same calyx, can also include several capsules in species such as the aconitum (figure 30) (§104). Depending on their shape, fruit clusters are described as ears, panicles, or umbels (§109). The question then arises whether such fruit clusters, bunches of grapes, or ears of wheat, constitute one or more fruits (§105–107). To settle the issue, Cesalpino argues that the

Figure 30: Plate of Aconitum (*Aconitum Napellus*).

unity of the fruit depends on the unity of the flower and the calyx; the calyx itself takes precedence over flowers in determining the individuality of a fruit (§108). Cesalpino may not have recognized the calyx as a true floral part,[574] but he did acknowledge it as a decisive floral and carpological trait.

Seeds without a fleshy pericarp are often contained in capsules, whose shape varies according to the species and may contain a single compartment (e.g., in roses) or several (e.g., in poppies) housing their seeds (§97). Cesalpino states that the seeds are sometimes accompanied by pappi, or down, like that of the rose (figure 31 and 32), allowing them to be spread by the wind (§98).

Figure 31: Picture of a ripe rose fruit/seed capsule section showing the down.

Figure 32: Picture of a rose seed capsule section showing the down.

574 According to Greene; see our commentary on chapter 7.

In §99, Cesalpino describes patterns of seed distribution in the capsules. Sometimes the seeds are in the center, as in the common corncockle (*Agrostemma gitagho*) (figure 33); in the "siliques" of leguminous plants, the seeds are on one side only; in the "pods" of the greater celandine, they are on both sides (figure 34). The other cases are not exemplified and remain unclear. They probably correspond to the different modes of embryo placentation within the fruit.

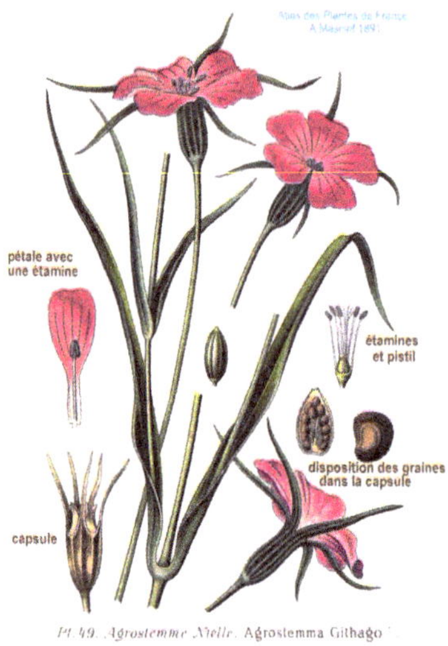

Figure 33: Plate of common corncockle (*Agrostemma gitagho*).

Figure 34: Plate of greater celandine (*Chelidonium majus* L.).

From today's botanical point of view, the differences outlined by Cesalpino are relevant, although the vocabulary has changed. Thus, he uses the term "capsule" for any type of organ that contains the seeds; this is not its current, more restricted technical sense. Similarly, his use of "silique" and "pod" is the reverse of current terminology, according to which the seeds are all attached on the same side in a pod and distributed on both sides in a silique (figure 35).

Figure 35: Left: Pod; Right Silique (current terminology).

In §100, Cesalpino mentions fruits with seeds on the periphery of the fruit, for example, strawberries, blackberries, or raspberries. Furthermore, capsules, pods, and siliques have a softer outer membrane and a harder inner membrane; Cesalpino explain this by resorting to his physiological theory of interlocking layers. Indeed, the capsules, siliques, and pods, as fruits, would have arisen from the bark, which itself has a softer outer layer (the cork, in current terminology) and a harder woody inner layer that he called the liber (the bast in current terminology) – as he posited in §32, §100, and §101. We propose below a summary diagram of Cesalpino's physiological theory of interlocking of layers, based on the anatomical descriptions of all the parts mentioned so far in *De Plantis* (figure 36).

Cesalpino then addresses the phenomenon of dehiscence of dry fruits, and explains how they expel their seeds (§101), unlike fruits whose seeds remain in the fleshy pericarp (squash, peaches, etc.). He identifies two parts of the fleshy pericarp: the outer layer of the fruit, and the softer fleshy part – in current terms, the epicarp (the skin of the fruit) and the mesocarp (§103) (figure 37).

Fruits are borne in various places on the plant depending on the species. Cesalpino describes cauliflory, i.e., the fact of bearing flowers on the trunk (§82, §111), and proposes the examples of the carob tree (*Ceratonia siliqua* L.) and of the Egyptian fig (*Ficus*

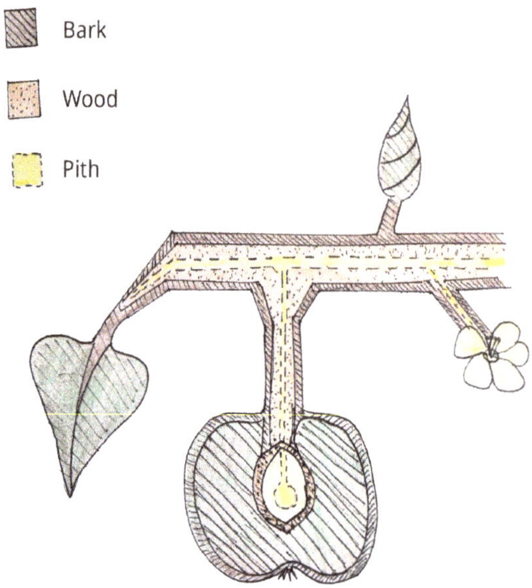

Figure 36: Diagram of Cesalpino's physiology theory of interlocking layers.

Sycomorus L.).[575] He notes that a plant, through its normal development, often flowers and bears fruit at successive moments of its growth and according to its age (§110, §111, §112). He also mentions herbaceous acaulescent plants, such as the violet, cyclamen, or mandrake, which bear each of their fruits at the end of a pedicel growing directly from the root (figure 38).

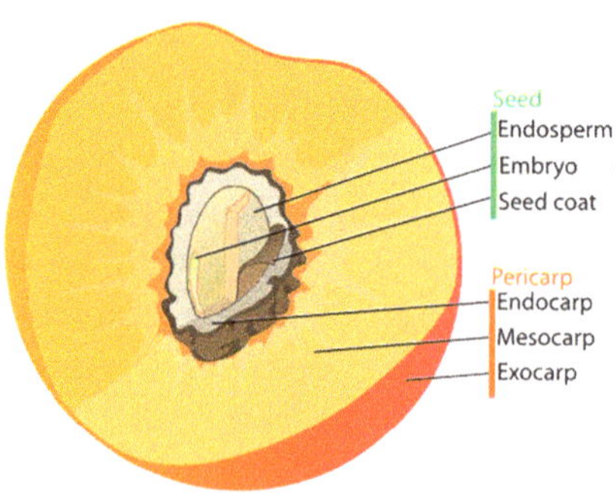

Figure 37: Diagram of the anatomy of a peach.

Figure 38: Picture of mandrake's fruits (*Mandragora officinarum* L.).

575 The Judas tree (*Cercis siliquastrum*), imported from the East at the time of the Crusades, is one of the best-known examples of cauliflory.

Chapter 11. The Secondary Parts (§114–117)

Chapter 11 deals with secondary plant parts such as the tendrils and hooks of climbing plants like grapevine and ivy. Plants use these parts to cling to a support "as if with hands" (*tanquam manibus*) to support their weight (§114). Cesalpino notes that tendrils are not located in the same place in all species; they may be found at the stipules of the leaves themselves, or on the side, in front, or at the tips of the leaves. For example, some leguminous plants (*Fabaceae*) such as pea (*Pisum sativum* L.) (figure 39) or vetch (*Vicia cracca*) (figure 40) develop their tendrils at the end of their leaflets, while grapevine (*Vitis vinifera* L.) develops tendrils in front of the leaves (figure 41). Cesalpino also observes that plants such as clematis use the "petioles" or "pedicels" (*pediculus*) of their leaves directly as tendrils (figure 42). He differentiates these types of climbing plants from ivy, which uses claws to cling to its supports (figure 43), and from voluble plants that wind their stems in a spiral around their support, like silkvine (figure 44) and, more commonly, bindweed and other *convolvulaceae* (figure 45).

Figure 39: Plate of pea (*Pisum sativum* L.).

Figure 40: Plate of vetch (*Vicia cracca*).

Figure 41: Picture of grapevine (*Vitis vinifera* L.) showing tendrils opposed to leaves.

Figure 42: Picture of clematis (*Clematis flammula* L.) showing petioles winding.

Figure 43: Picture of ivy (*Hedera helix* L.) showing its claws.

Figure 44: Picture of silkvine (*Periploca graeca* L.) stem around a tree.

Figure 45: Picture of bindweed (*Convolvulus arvensis*) stem.

These observations prompt him to investigate how plants climb, and he wonders whether climbing plants:

> quasi sensus quidam adiacentis corporis illis videatur inesse, cum repant, donec inueniant, et inuentum apprehendant.

> possessed a kind of sensation of the body next to them, since they creep until they find it, and having found it, grab onto it (§114).

Cesalpino has noticed that creepers grope around for adjacent objects, which seems to imply that they possess the sense of touch, and thus sensation,[576] even though he explicitly referred in §2 to the Aristotelian doctrine of the absence of sensation in plants. Yet, here, he seems to surpass Aristotelianism, as he lends purpose and intentionality to plants that crawl in search of a support, and "grab onto it" (*apprehendant*).

This evocation of sensation in plants is perhaps not anecdotal. Chapter 2 discussed the problem of plant nutrition and failed to explain how plants select and absorb food although they do not possess sensations, since none of the physical analogies proposed by Cesalpino were satisfactory. Moreover, Cesalpino must have been influenced by a passage from Pseudo-Aristotle's *De plantis* stating that everything that feeds itself, including plants, has sensations:

> Possumus etiam alio modo dicere quod est animatum est non dico tamen quod sit inanimatum, si habuit animam et aliquem sensum, quia res quae cibatur non est sine anima.
>
> We can say in a different way that it [i.e., the plant] is animate – and I do not say that it is inanimate – since it has a soul and some kind of sensation, for a thing that is nourished is not without a soul.[577]

Even though this passage is probably an apocryphal addition from the Arabic tradition, which is not truly consistent with the rest of the treatise, nor with Aristotelianism in general,[578] it should not be overlooked. Indeed, Albert the Great had already discussed – in *De vegetabilibus*, one of Cesalpino's probable sources – the existence of a sense of touch and taste of plants, relying on Pseudo-Aristotle's *De plantis* to a large extent,[579] and contemporaries of Cesalpino also used this treatise as a source for (re)theorizing an Aristotelian botany.[580] Cesalpino's very choice of *De plantis* as the title of his own treatise *De plantis libri XVI* should be regarded more as a reference than as a coincidence. Cesalpino was also familiar with *De causis plantarum* by Theophrastus who, although

576 Ogilvie (2006: 224) also noticed this idea in Cesalpino, but only mentioned it in passing.

577 Nicolaus Damascenus [Pseudo-Aristotle], *De plantis* §27, 816b4–6 (translation adapted from the ed. and French transl. Federspiel and Cronier 2018: 92).

578 On this point, see Droossaart Lulofs-Poortman 1989: 261.

579 Albert the Great, *De vegetalibus* I, 1, 10, §67–68 (ed. Meyer 1867: 36–37). Albert the Great conclusion is that the apparent sensations and desire of plants are of another kind than those of animals, and that from an aristotelian point of view, there is no real perception and desire in plants because these two faculties cannot exist through the vegetative soul and the material body only (Panarelli 2021: 100–101). For an in-depth discussion of the issue see: Panarelli 2020: 137–162; Cerrito 2021: 105–122. See also Meyer 1854–1857: 40–42.

580 For example, Julius Caesar Scaliger, who contested that this text was written by Aristotle (Scaliger 1556), or Andrés Laguna de Segovia (1499–1559) who published a Latin printed edition and commentary of the text (Laguna de Segovia 1543). See Baldassarri 2020: 237–264. According to Baldassarri, the influence of the Pseudo-Aristotle's *De plantis* on Renaissance botanists was, however, weak or even absent. Gaspard Bauhin (1623: 493, 498) and Adam Zalužanský (1592) were exceptions. Scholars and philosophers interested in the vegetative soul discussed more the text.

he adhered to Aristotelianism, seems to attribute some sensation to plants, at least in one passage. He presents plant sensation as a mediating operator between the plant and the physical causes influencing its reversible movements:

> Ἡ δὲ τῶν ἀνθῶν σύμμυσις καὶ δίοιξις ἐλαφροτέρα καὶ ῥᾴων ἰδεῖν· ὑπὸ γὰρ τοῦ ψύχους καὶ τῆς ἀλέας γίνεται ψυχρῶν ὄντων καὶ ἀσθενῶν· συμμύει μὲν γὰρ ξυνιόντος καὶ οἷον πηγνυμένου τοῦ ὑγροῦ, συναπολείπει γὰρ καὶ τὸ θερμόν, ἀνοίγεται δὲ πάλιν διαχεομένου καὶ ἀνιέντος ὅπερ ὁ ἥλιος ποιεῖ. Τὰ δὲ πλέον καταδυόμενα καὶ ὑπερίσχοντα δῆλον ὅτι ψυχρότερα καὶ ἀσθενέστερα δι' ὃ μᾶλλον συμπάσχει ταῖς μεταβολαῖς. Ἡ δὲ αἴσθησις οὕτως ὀξεῖα γινομένη τοῖς καθ' ὕδατος οὐκ ἄλογος ἄλλως τε καὶ ἐν τόποις θερμοῖς καὶ ἐμπύροις.
>
> The closing and opening of the flowers is a less difficult matter and easier to solve, since it is brought about by cold and heat, the flowers being cold and weak. Thus, they close up when their fluid condenses and (as it were) freezes (since at this time their heat leaves them too), and open when the fluid dissolves again and thaws, this being done by the sun. The plants that sink under the water and emerge above it to a greater extent are evidently colder and weaker than the flowers, and for this reason more affected by the changes. That a plant under water should be so keenly sensitive is not unreasonable, especially in a torrid region of fiery heat.[581]

Although it contradicts *De anima*, the attribution of touch and taste to plants would allow Cesalpino to account for the observation of movement in (climbing) plants, since vital movement results from desire and is directed by sensation, and to solve, at least partly, the problem of plant nutrition.[582]

This analogy with animal behavior conflicts with the conception of the chain of beings derived from Aristotelianism, but Cesalpino's reflections on this subject confirm that his sense of observation can induce the botanist in him to suggest ideas that are incompatible with the doctrine that he endorses as a philosopher. This emancipation of scientific thought from philosophy (and from religious doctrines) is typical of the transition towards early modern experimental science, and is also reflected, for example, in Galileo's criticism of Aristotle's physics. Cesalpino does not, however, wholly discard Aristotle's theory of the soul, sensations, and the living. Aristotelian psychology is a framework for Cesalpino's theoretical reflection on the faculties of plants but does not prevent the Italian botanist – perhaps unintentionally – from pointing out Aristotle's limitations or inconsistencies. On the question of plant sensitivity, 21st-century biology converges with Cesalpino's intuitions, since we now know that all plants, and more particularly climbing plants, can discriminate, and react to, tactile stimuli.[583]

After this allusion to plant sensation, Cesalpino describes the thorns of some species and their defensive role (§115). He then discusses parts that can be described as "acci-

581 Theophrastus, *De causis plantarum* II, 19, 3, 6 – 4, 5 (ed. and Fr. transl. Amigues 2012–2017: 1.114; Engl. transl. Einarson and Link 1976–1990: 1.357–359).

582 This is developed further in Hiernaux and Tresnie 2023a.

583 Trewavas 2014: 10–19.

dental" (*ob solam necessitatem additae*), in that they are not present in all members of a species, but only in some individuals, and apparently have no function (*nullius gratia data*) for the plant – although they might be of use to humans. For example, the trunks of older trees may be covered by mosses or lichens, and galls and growths may develop in individuals depending on their species (§116) (figure 47 and 48). In this respect, Cesalpino's position is perhaps closer to that of Theophrastus, who considers moss (βρύον) as an annual part of the tree,[584] than to that of Aristotle, who sees a parasitic plant such as mistletoe as a true plant generated by spontaneous generation.[585]

However, what Cesalpino describes as a "beard" (*iuba*) hanging from the branches of some trees could be epiphytic plants or lichens such as the old man's beards (figure 46). Similarly, mosses and lichens in general are not parts of the tree, but other plant species.

Figure 46: Picture of old man's beards (*Usnea barbata*).

584 Theophrastus, *Historia plantarum* I, 1, 2 (ed. and Fr. transl. Amigues 1988–2006: 1.2–3; ed. and Engl. transl. Hort 1916–1928: 1.3–5).

585 Aristotle, *De generatione animalium* I, 1, 715b25–30 (ed. and Fr. transl. Louis 1961: 3; ed. and Engl. transl. Peck 1943: 9).

Figure 47: Picture of a *Quercus coccifera* L. gall.

Figure 48: Picture of the dog rose gall or Bedeguar.

Chapter 12. The Four Main Genera and their Divisions (§118–134)

Now that Cesalpino has described the main organs of plants, he can use their morphology to propose his own method of classification. He recapitulates the four main "genera" (*genera*) of Theophrastus, namely "trees" (*Arbor*), "shrubs" (*Frutex*), "undershrubs" (*Suffrutex*) and "herbs" (*Herba*) (§118), but criticizes Theophrastus for following popular usage and admitting taxonomic groups based on plant habitat, or on practical

applications.[586] With his method of classification, Cesalpino returns to the theoretical core of Aristotelian psychology, rather than following Aristotle himself, as well as Theophrastus and Dioscorides, in their concessions to common usage.

Cesalpino identifies a hierarchy of at least four taxonomic groups: Theophrastus' four main genera; their division in lower groups (subgenera of the four main genera, or families); genera in the proper sense (group of species); species (as group of individuals). However, his vocabulary is imprecise, and he qualifies as genera all groupings above species. Indeed, it was not until 1689 that French botanist Pierre Magnol (1638–1715) introduced the term "family" to describe a grouping of related genera.[587] Tournefort is generally recognized for his clear and explicit definition of the concept of genus as a group of related species at the very end of the 17th century.[588] Nevertheless, *De plantis*, and other contemporary botanical works by Ghini, Gessner, or Bauhin, for example, reveal that these taxonomic concepts, although not explicitly defined, were already in use in the 16th century.[589] Tournefort writes in his *Élements de botanique*:

> Voyant donc qu'il restait beaucoup à faire sur le traité des genres [...] Mon dessein fut d'abord de me convaincre une fois en ma vie s'il était possible de trouver des règles assurées pour établir les genres, et les classes des plantes; et comme il me parut que je retombais de moi-même dans les vues de Gesner, de Césalpin et de Columna, je crus qu'il était bon de faire voir les avantages qu'on pouvait tirer pour les éclaircissements de la Botanique.
>
> Seeing therefore that there was still much to be done on the treaty of genera [...] My first intention was to convince myself once in my life whether it was possible to find assured rules for establishing the genera, and classes of plants; and as it seemed to me that I was falling back on the views of Gessner, Cesalpino and Colonna, I thought it was good to show the advantages that could be drawn for the clarification of Botany.[590]

In the classification adopted in *De plantis*, Cesalpino groups trees and shrubs together on account of their wood, their large size, and their perennial life expectancy. He distinguishes them by the number of their stems: a single trunk for trees, several for shrubs (§119). Undershrubs and herbs are made of softer material, although the former can live for several years. Herbs are annuals (sometimes bi- or tri-annuals) and not perennials, as they perish after the production of their seeds (§122). They are also characterized by the growth of leaves from the root (§124).

586 Compare ch. 12: §130–134 with Theophrastus, *Historia plantarum* I, 3, 1 (ed. and Fr. transl. Amigues 1988–2006: 1.9–10; ed.and Engl. transl. Hort 1916–1928: 1.23–25).

587 Magnol 1689. See Magnin-Gonze 2009: 127.

588 Tournefort 1694: 17–39.

589 The few comments on plants left by Ghini (Cesalpino's mentor) show he had already engaged in identifying several species conflated under a single (genus) name (Morton 1981a: 156n40).

590 Tournefort 1694: 20, our translation. In these pages, Tournefort also acknowledges Gaspard Bauhin for his conception of genera.

Cesalpino acknowledges the need for more precise subdivisions of the four main genera (families and genera in current terminology) to allow the multiplicity of species to be classified; he will create these subdivisions, as "few are known for the moment" (*pauca adhuc nota sunt*) (§126). Cesalpino first rejects the genera classification derived from food and medicinal uses as unreliable (§127, §129), and criticizes subjective criteria such as taste and smell to classify plants;[591] such criteria, as well as superficial resemblances, are common to many morphologically unrelated plants. Some morphological secondary characteristics, such as the bearing of thorns, are also unhelpful primary criteria for classification (§128), because these characteristics are observed across the four main genera. This indicates that Cesalpino is looking for criteria of division that would permit the creation of a coherent vertical hierarchy of taxa.

Cesalpino presents his thoughts on how to define a genus from the study of similarities between plants. Similarities observed in the form or matter of roots, stems, leaves, or colors of the flower are unsuitable, because they allow transversal, artificial groupings across the four main genera (§131, §132). These characters are excluded from this section, and therefore should not be prioritized to operate an – arbitrary – division of the four main genera. They are less decisive than roots, stems, and then fruits and seeds; in the following chapter, Cesalpino justifies rationally that these characters are the most important, according to the natural order of functions, which is reflected in plants parts. However, Bremekamp shows that Cesalpino considered all morphological characters in latter stages of his classification, including roots, stems, and flowers.[592]

Cesalpino sets out his own analysis of the classification problem in §134, a very important section, although it is sometimes ambiguous. He suggests that a large number of similarities alone is not sufficient to establish genera; the general comparison of a multiplicity of similarities is only relevant in the last stage of the analysis, to compare and group together very similar species. A more reliable method and more specific criteria must therefore have been employed to identify that these species belonged together. Indeed, not all similarities are equally relevant. The division of the main genera, leading to subgenera (i.e., families), must therefore be founded on precise and clearly defined morphological criteria: specific, exclusive differences, hierarchically ordered. Such criteria are difficult to determine, which is why Theophrastus and Dioscorides adopted utilitarian, medicinal, or even habitat criteria (§135). Cesalpino criticizes these options as easy solutions resulting in artificial classifications.

From the point of view of the history of science, Cesalpino is laying the foundations of a method founded on the examination of all the characters of plants, and on their hierarchization, a method resulting in the development of a natural classification system. On the contrary, artificial systems of classification rely on the examination of

591 Ancient botanical tradition, at least since Theophrastus, valued the description of the smells and tastes of plants, to which Theophrastus devotes the twenty chapters of Book VI of *De causis plantarum*.
592 Bremekamp 1952.

only one character – or very few – deemed to be decisive; for instance, the classification adopted by de L'Obel relied on the leaf, and Linnaeus' sexual system was based solely on the number and arrangement of pistils and stamens. In §132–134, Cesalpino explain, through the example of many species, that these systems are "artificial" because they can result in counter-intuitive groupings or separations. He also suggests that, although all morphological characters may be considered, they are not equally relevant, as Adanson later asserted. Cesalpino is undoubtedly the first botanist to promote a form of weighing of characters, when he recommends giving priority to characters that enable the elaboration of higher, intermediate, and specific taxonomic groups, Cesalpino is undoubtedly the first botanist to promote a sort of weighing of characters. This weighing of characters according to their degree of stability among taxonomic groups for establishing a hierarchical classification is the key principle of the natural method, as will be explained and demonstrated by Antoine-Laurent de Jussieu (1748–1736).[593] However, Cesalpino's incomplete attempt to weigh characters according to their importance for classification is not logically accompanied by a methodological order of priority for applying the division and determining species.[594]

Thus, Cesalpino perceived the crucial methodological elements to reach a genuine natural classification, when he suggested that all characters should be examined before effecting hierarchical divisions, and that these divisions should be determined by specific differences exclusive to each group, and when he undertook to weigh characters and determine their relative priority. However, Cesalpino's hierarchical classification is not presented as purely inductive, although it was probably elaborated by induction. Rather, it is justified by metaphysical considerations about the faculties of the soul and their respective importance, according to essentialism and finalism (§137). Indeed, Cesalpino considers roots and stems as the most relevant characters to divide the main genera because they correspond to the first operation of the vegetative soul – nutrition. Fruits and seeds come second, because they correspond to the second operation of the soul – reproduction.[595] The non empirical nature of this rationale may have been the reason why some commentators considered Cesalpino's classification "artificial".[596] And yet, both his method and the results that he obtained are "natural"; his rationale only may be deemed artificial. There are separate epistemological issues, and they should not be conflated.

593 Jussieu 1789. On Jussieu, see Stevens 1994; Magnin-Gonze 2009: 150–156.

594 Atran 1990: 156.

595 Tournefort (1694: 6) was one of the first to criticize this finalist way of ordering classification characters in botany (Leroy 1957: 191).

596 For example, Sachs, Harvey-Gibson, and Arber; see the section "Positioning of Cesalpino in the history of classification", in part I.

Chapter 13. Criteria for Defining Subgenera and Species (§135–144)

In this chapter, Cesalpino examines how to choose effective classification criteria. He argues that the substance of a species cannot be identified from its accidents – an Aristotelian stance –, and refers to his *Quaestiones peripateticae* (§135). He then diverges from Aristotelianism, because Aristotelian principles used for the classification of animals are not applicable to plants: animals are grouped according to the manner in which the sensitive faculty of their soul (their form) is expressed. Indeed, they do not all possess the same type of sensations, and therefore present a variety of sense organs and of organs of locomotion (§143). On the contrary, Cesalpino stresses that the vegetative faculty of the souls of plants expresses itself in the same way in all plants – this is also true for the intellective faculty of the human soul, which defines a single species (§143). And yet, plants, unlike humans, are not a single species, because they differ in the matter (the organs) through which their vegetative faculty is expressed (unlike the intellective faculty of humans, which does not depend on any physical organ). Therefore, the differences observable in plant parts offer the best criteria for dividing plants into genera and species (§136).

Cesalpino justifies philosophically the choice of the first criterion selected for dividing plants into the four main genera. The primary operation of the plant soul is nutrition, and nutrition depends on roots and stems; therefore, the structures of roots and stems – woody or not – must be given priority in the classification. The next criterion of division is the number of stems (§138). The intersection of the two criteria is represented in the table below (table 2):

Table 2: Table of the four main genera.

	Single stem	**Multiple stems**
Woody (Wood)	Trees	Shrubs
Non-woody	Herbs	Undershrubs

Next, Cesalpino proposes to subdivide the four main genera according to the other characteristics of stems and roots: size, shape, number, etc. At this point, his method of division no longer appears dichotomous. These characteristics of stems and roots take precedence over those of leaves (§123), because leaves intervene in the second operation of the soul, i.e., reproduction (leaves protect the fruit, according to Cesalpino). The choice and weighing of the criteria are not justified by the observation of greater empirical stability or reliability, rather, they result from an aprioristic, metaphysical interpretation. And yet, the observation that this weighing of plant characters produced a more reliable classification may have led Cesalpino to seek a philosophical justification in Aristotelianism, and to present this justification, a posteriori, as principial in

the theoretical part of *De plantis*.[597] Linnaeus also gives priority to theoretical considerations over preliminary observation.[598] Aristotelian argumentation does not prove that Cesalpino's method of classification was purely deductive;[599] he had already developed his classification system in his 1563 herbarium and had been using it for at least twenty years.

The next level of division (i.e., the division of subgenera/families in genera) relies on fruiting, and the parts that make it possible (leaves, flowers, and fruits). The fruit is a privileged criterion of classification, because:

> In nullis enim aliis partibus tantam organorum multitudinem et distinctionem natura molita est, quanta in fructibus condendis spectatur.
>
> Nature has endeavored to produce as much diversity and difference in no other part than that which we see in the development of the fruit (§142).

The fruit is the ultimate achievement of plants (§143). At this step of division, a teleological justification is again combined to the inductive, meticulous observation of the great diversity and stability of fruits and seeds characteristics, which results in the identification of genera and species. Cesalpino notes that not all the categories defined in this way are necessarily named. This indicates that his innovative method allows him systematically to identify new genera that had not previously been grouped by usage but are now grouped for morphological reasons (§141). Cesalpino's method is also predictive and enables the classification of yet unknown species.

Since plant soul only performs two operations, there are only two hierarchical levels of division of the main genera. These divisions are not dichotomous, and therefore result in a multiplicity of genera located on the same level. The main categories of Cesalpino's classification tree can be schematized as follows:

Living
Plants – animals
Main genera: Herbs – undershrubs – shrubs – trees
Subgenera (stems and roots) (i.e., "families")
Genera and Species (flowers, leaves, and fruits).[600]

597 Mayr 1982: 160.
598 Staffleu 1971: 42.
599 See our introduction and the section "Positioning of Cesalpino in the History of Classification" in Part I for a broader discussion of this topic with references.
600 For a detailed analysis, see Bremekamp 1952 and figures 5, 6 and 7 of this book.

Chapter 14. Subdivisions Based on the Reproductive Operation (§145–157)

Cesalpino has posited that fruits allow for the most accurate classification of genera into species. However, some plants are imperfect in that they do not bear fruit: for example, lichens, fungi, algae, and corals – our understanding of Cesalpino's "marine shrubs" (*frutex mari*). He suggests that these *plants* use the first operation of the soul to compensate for the absence of the second, which explains why they feed and grow proportionally more than plants that bear fruit (§145).

This remark is highly questionable as a classification criterion, and yet, the distinction between flowering/seed plants (Spermatophytes and Angiosperms) and non-flowering/seedless plants (Cryptogams) is fundamental, and still valid, in modern plant taxonomy. Fungi and other organisms mentioned by Cesalpino are no longer included in the plant kingdom.

Cesalpino then studies a group of plants that have inflorescences but (according to him) no seeds, for example, the broomrapes, the cytinus, the orchids or certain cereals whose ears are empty (§146). The imperfection of this group of plants is congenital and must be distinguished from the contingent sterility of individual plants, considered more perfect (§147). Cesalpino therefore includes this among specific differences.

The structure of orchids, broomrapes, and cytinus is such that their tiny seeds are invisible to the naked eye. Without a microscope, Cesalpino would only have been able to see the fine dust or mucus made of these seeds.

He then identifies plants whose means of reproduction are analogous to seeds, but are not seeds, but spores. He observes these spores under the leaves of various species of ferns, which he groups together separately (§148). What he describes as "a sort of down" (*lanugo quaedam*) on the leaves of ferns probably refers to their sporangia (figure 49).

Unlike his incomplete analysis of orchid reproduction, this observation of ferns reflects the current botanical distinction between spermatophytes (plants with seeds) and pteridophytes (vascular plants without seeds, which reproduce with apparent spores, like ferns).

In the next paragraph, Cesalpino proposes to divide the vast group of seed plants, considered more perfect than the plants belonging to the three previous groups, according to their morphology. For this purpose, he examines the "number", "position" (or "location"), and "shape" (*numerus, situs, et figura*) of the organs (§149). These three criteria are sufficient to differentiate the fruits: the other characteristics of fruits are either derived from these very criteria (e.g. size can be analyzed as a "shape"; cohesion/fusion of parts as "number"), or are secondary and therefore not prioritized for classification (e.g., color or texture of parts) (§150).[601] Cesalpino's statement may indicate that

601 According to Atran (1990: 152): "Direct functional value pertains to the sensible realm exclusively. This is not Cesalpino's fundamental preoccupation; rather, he is concerned with the *rationality* imposed

Figure 49: Sporangia under a leaf of *Asplenium scolopendrium* L.

the criteria should be applied in the order in which they are listed: first number, then position, then shape. This method of division is also applicable to the groups of plants belonging to the four main genera (§153).

The criterion of number allows the differentiation between flowers that produce a single seed or a single seed receptacle, and flowers that produce two, three, four or more (§151).

The criterion of location creates a distinction between plants whose seed receptacles are below, or above the flowers (in current terminology, the distinction between inferior and superior ovaries), or whose flower is below, or around the fruit (hypogynous and epigynous flowers) (§152). The position of the flower in relation to the fruit is explicitly described in the text, but that of the capsule/ovary is less clear. However, since the implantation of the flower reciprocally conditions that of the fruit, the position of

by God on the material being. That is why only the number (*numerus*), position (*situs*) and figure (*figura*) of the seeds and other selected floral organs are truly essential to taxonomy." The theological interpretation of Cesalpino's taxonomy is invalidated by the absence of God in *De plantis*, and by Cesalpino's assertion that other properties vary according to the number, position, and shapes of plant parts – an assertion that must result from observation at least as much as from rationality.

the seeds, in Cesalpino's description, likely corresponds to the inferior or superior position of the ovary (figure 19).

Finally, the criterion of shape applies to flowers, receptacles, or seeds. The texture of these organs provides a secondary criterion (§153).

According to Cesalpino, the characteristics of the fruits are the most relevant to the classification of species, although properties of the stems and leaves can also be significant, since they play an indirect (or accidental) part in fruiting (§154). Cesalpino adds that a plant character can only constitute an effective classification criterion if it is stable – characters that vary according to geography, climate, or method of cultivation should be avoided. These variations allow at best the division of a species into different varieties, but are irrelevant for classification, as are colors, odors or vapors (§155), tastes, and medicinal properties (§156). In Cesalpino's words, *per se* differences manifest themselves "everywhere in the same way" (*vbique eodem modo*). His attempt to build a natural method of classification based on stable criteria suggests, again, the inductive nature of his study.

Now that he has outlined his classification method, Cesalpino announces the plan of the following books, in which he organizes and describes the species according to his selected criteria.[602]

602 For more details on the classification of books II to XVI, see Bremekamp 1952.

Part IV: **Appendices**

https://doi.org/10.1515/9783111001104-004

Bibliography

Primary Sources

Edited sources

Albert the Great, *De vegetabilibus*: Ernst Meyer (ed.), *De vegetabilibus libri VII*. Berlin: Reimer, 1867.

Aristoteles, *Analytica*: William David Ross (ed.), *Aristotle's Prior and Posterior Analytics*. Oxford: Clarendon Press, 1949.

Aristoteles, *De anima*: William David Ross (ed.), *Aristotle. De anima, with introduction and commentary*. Oxford: Clarendon Press, 1961.

Aristoteles, *De Generatione animalium*:

Pierre Louis (ed. and Fr. transl.), *Aristote. De la génération des animaux* (Collection des Universités de France). Paris: Belles Lettres, 1961.

Arthur L. Peck (ed. and Engl. transl.), *Aristotle XIII. Generation of Animals* (Loeb Library 366). Cambridge, MA : Harvard University Press, and London : William Heinemann, 1943.

Aristoteles, *De Incessu animalium*: Andrea Falcon and Stasinos Stavrianes (ed.), *Aristotle on How Animals Move: The De incessu animalium: Text, Translation, and Interpretative Essays*. Cambridge: Cambridge University Press, 2021.

Aristoteles, *De Motu animalium*: Pierre Louis (ed.), *Aristote. Marche des animaux. Mouvement des animaux. Index des traités biologiques* (Collection des Universités de France). Paris: Belles Lettres, 1973.

Aristoteles, *De partibus animalium*:

Pierre Louis (ed. and Fr. transl.), *Aristote. Les parties des animaux* (Collection des Universités de France). Paris: Belles Lettres, 1956.

Arthur L. Peck (ed. and Engl. Transl.), *Aristotle. Parts of Animals. Movement of Animals. Progression of Animals* (Loeb Classical Library 323), Cambridge, MA: Harvard University Press, and London: William Heinemann, 1945.

Aristoteles, *Ethica Nicomachea*: I. Bywater (ed.), *Aristoteles Ethica Nicomachea.* Oxford: Oxford University Press, 1894.

Aristoteles, *Historia animalium*: Pierre Louis (ed.), *Aristote. Histoire des animaux* (Collection des Universités de France), 3 vols. Paris: Belles Lettres, 1964–1969. (Engl. transl. d'Arcy Wentworth Thompson, *The Works of Aristotle*, volume IV: *Historia Animalium.* Oxford: Clarendon Press, 1910).

Arisoteles, *Metaphysica*: Werner Jaeger (ed.), *Aristotelis Metaphysica.* Oxford: Oxford University Press, 1957.

Aristoteles, *Parva naturalia* [including *De juventute et senectute, De vita et morte, De respiratione*]: William David Ross (ed.), *Aristotle. Parva Naturalia, a revised text with introduction and commentary*. Oxford: Clarendon Press, 1955.

Aristoteles, *Physica*: William David Ross (ed.), *Aristotle. Physics, a revised text with introduction and commentary*. Oxford: Clarendon Press, 1936. (Engl. transl. Charles David Chanel Reeve, *Aristotle. Physics.* Indianapolis: Hackett, 2018).

Dioscorides, *De materia medica*: Max Wellmann (ed.), *Pedanii Dioscuridis Anazarbei de materia medica libri quinque*, 3 vols. Berlin: Weidmann, 1906–1914 (reprints: 1958 and 2000) (Engl. transl. Lily Beck, *Pedanius Dioscorides of Anazarbus. De materia medica* (Altertumwissenchaftliche Texte und Studien 38). Hildesheim, Zürich and New York: Olms, 2017.

Herodotus, *The Histories*: George Rawlinson (Engl. transl.), *The history of Herodotus with copious notes and appendices, illustrating the history and geography of Herodotus, from the most recent sources of information; and embodying the chief results, historical and ethnographical, which have been obtained in the progress of cuneiform and hieroglyphical discovery*, 4 vols. London: J. Murray, 1858–1860.

Hippocrates, *On generation. On the nature of the child. Disease IV:* Iain Malcolm Lonie (ed.), *The Hippocratic treatises. On generation, On the nature of the child, Disease IV: A Commentary* (Ars Medica, Texte und Untersuchungen zur Quellenkunde der Alten Medizin, II.7). Berlin and New York, NY: De Gruyter, 1981.

Pliny the Elder, *Naturalis historia*:

Jacques André, Jean Beaujeu, Raymond Bloch, Jean-Michel Croisille, Eugène De Saint-Denis, Jehan Desanges, Alfred Ernout, Jean Filliozat, Henri Le Bonniec, Roger Pépin, Robert Schilling, Guy Serbat, Alain Silberman and Hubert Zehnacker (eds and Fr. transl.), *Pline l'Ancien. Histoire Naturelle* (Collection des Universités de France), 37 vols. Paris: Belles Lettres, 1947–2015.

Alfred Carleton Andrews, David Edward Eichholz, William Henry Samuel Jones and Harris Rackham (eds and Engl. Transl.), *Natural History* (Loeb Classical Library 330, 352–353, 370–371, 392–394 & 418–419), 10 vols. Cambridge, MA: Harvard University Press, 1938–1962.

Nicolaus Damascenus, *De plantis*:

Hendrik Joan Drossaart Lulofs and Evert Lubbertus Jacobus Poortman (eds and Engl. transl.), *De plantis. Five translations* (Aristoteles Semitico-Latinus). Amsterdam, Oxford and New York: North Holland Publishing Company, 1989.

Michel Federspiel and Marie Cronier (Fr. transl.), *Pseudo-Aristote. Du monde; Positions et dénominations des vents; Des plantes* (La roue à livres). Paris: Les Belles Lettres, 2018, pp. 87–127.

Theophrastus, *Historia plantarum*:

Suzanne Amigues (ed. and Fr. transl.), *Théophraste. Recherches sur les plantes* (Collection des Universités de France), 5 vols. Paris: Belles Lettres, 1988–2006.

Arthur Hort (ed. and Engl. Transl.), *Enquiry into Plants and Minor Works on Odours and Weather Signs* (Loeb Classical Library 70 & 79), 2 vols. London: W. Heinemann, 1916–1928.

Theophrastus, *De causis plantarum*:

Suzanne Amigues (ed. and Fr. transl.), *Théophraste. Les causes des phénomènes végétaux* (Collection des Universités de France), 3 vols. Paris: Belles lettres, 2012–2017.

Benedict Einarson and George Konrad Karl Link (ed. and Engl. transl.), *De causis plantarum* (Loeb Classical Library 471 & 474–475), 3 vols. Cambridge, MA: Harvard University Press, 1976–1990.

Varro, *De re rustica*: Jacques Heurgon and Charles Guiraud (ed. and Fr. transl.), *Varron. Économie rurale* (Collection des Universités de France), 3 vols. Paris: Belles Lettres, 1978–1997.

Early printed sources

Adriani 1518: Marcello Virgilio Adriani, *Pedacii Dioscoridae Anazarbei de medica materia libri sex. Interprete Marcello Virgilio Secretario Florentino: Cum eiusdem annotationibus*, 6 vols. Florentia: Iunta, 1518.

Alpini 1592: Prospero Alpini, *De plantis Aegypti liber.* Venitii: F. de Franceschi di Siena, 1592.

Anguillara 1561: Luigi Anguillara, *Semplici, dell' eccelente Luigi Anguillara, li quali in piu pareri a diversi nobili huomini scritti appaiono, et nuovamente da M. Giovanni Marinello mandati in luce.* Venetiis: Vincenzo Valgrisi, 1561.

Barbaro 1492: Ermolao Barbaro, *Castigationes Plinianae Hermolai Barbari Aquileinsis Pontificis.* Roma: Eucharius Silber, 1492.

Barbaro 1516: Ermolao Barbaro, *In Dioscoridem ... Tralatum annotamenta. Quibus morborum et remediorum vocabula obscuriora in usum etiam mediocriter eruditorum explicantur (= Corollarium Dioscorides).* Venetiis: Johannes Baptista Egnatius (= Cipelli), 1516.

Bauhin 1623: Caspar Bauhin, *Pinax theatri botanici, sive index in Theophrasti, Dioscoridis, Plinii et botanicorum qui a seculo scripserunt.* Basel: Ludwig König, 1623.

Brasavola 1536: Antonio Musa Brasavola, *Examen omnium simplicium medicamentorum quorum usus est publicis officinis.* Romae: Bladus, 1536.

Brunfels 1530–1536: Otto Brunfels, *Herbarum vivae eicones ad natural imitationem*, 3 vols. Argentoratum: Johann Schott, 1530–1536.

Camerarius 1694: Rudolf Jakob Camerarius, *Ad Thessalum D. Mich. Bernardum Valentini professorem giessensem excelentissimum. De sexu plantarum epistola*. Tübingen: Romincius, 1694.

Candolle 1824–1873: Augustin-Pyramus de Candolle, *Prodromus systemati naturalis regni vegetabilis, sive, enumeratio contracta ordinum generum specierumque plantarum huc usque cognitarum, juxta methodi naturalis normas digesta*, 17 vols. Paris: Sumptibus Sociorum Treuttel et Würtz, 1824–1873.

Candolle 1832: Augustin-Pyramus de Candolle, *Physiologie végétale, ou exposition des forces et des fonctions vitales des végétaux*. Paris: Béchet Jeune, 1832.

Cesalpino 1571: Andrea Cesalpino, *Peripateticarum quaestionum libri quinque ad potentissimum et foelicissimum Franciscum Medicem Florentiae et Senarum Principem*. Venetiis: apud Iuntas, 1571.

Cesalpino 1580: Andrea Cesalpino, *Daemonum investigatio peripaterica. In que explicatur locus Hippocratis in Progn. Si quid diuinym in morbus habetur.* Florentia: apud Iuntas, 1580.

Cesalpino 1583: Andrea Cesalpino, *De plantis libri XVI ad Serenissimum Franciscum Medicem, Magnum Aetruriae Ducem*. Florentia: apud Georgium Marescottum, 1583.

Cesalpino 1593: Andrea Cesalpino, *Quaestionum peripateticarum lib. V. Ad Sereniss. Franciscum Medicem Magnum Hetruriae Ducem II. Daemonum inuestigatio peripatetica. Ad illustriss. ac Reverendiss. Archiepiscopum Pisanum Petrum Iacobum Borbonium ex Marchionibus Sanctae Mariae. Secunda editio. Quaestionum medicarum libri II. De medicament. facultatibus lib. II. Ad Sereniss. Ferdinandum Medicem Magnum Hetruriae Ducem III. Nunc primum editi.* Venitiis: apud Iuntas, 1593.

Cesalpino 1596: Andrea Cesalpino, *De metallicis libri tres ad Sanctissimum Dominum Nostrum Clementem VIII Pont. Max.* Romae: ex typographia Aloysij Zannetti, 1596.

Cesalpino 1602: Andrea Cesalpino, *Ars medica*. Romae: Aloysium Zannettum, 1602.

Cesalpino 1606: Andrea Cesalpino, *Praxis universae artis medicae, generalium aeque, ac particularium humani corporis praeter naturam affectuum dignotionem, iuditium et curam omnium uberrume complectens, summon labore, et studio concinnata, et unum recenter in volume coniecta. Perillustru et summe excellenti viro Fabio Pacio dicata.* Tarvisii: sumptibus Roberti Meietti, 1606.

Colonna 1592: Fabio Colonna, *Φυτοβασανος sive plantarum aliquot historia in qua describuntur diversi generis plantae veriores, ac magis facie, viribusque respondentes antiquorum Theophrasti, Dioscoridis, Plinij, Galeni, aliorumque delineationibus, ab alijs hucusque non animaduersae. ... Accessit etiam piscium aliquot, plantarumque nouarum historia. Ad illustrissimum et reverendissimum D. S. R. E Card. M. Antonium Columnam.* Neapoli: apud Io. Iacobum Carlinum et Antonium Pacem, 1592.

Colonna 1649: Fabio Colonna, *Rerum medicarum novae Hispaniae thesaurus seu plantarum animalium mineralium mexicanorum historia. Ex Francisci Hernandez novi orbis medici primarij relationibus in ipsa mexicana urbe conscriptis a Nardo Antonio Reccho ... collecta ac in ordinem digesta... A Ioanne Terrentio ... notis illustrata. ... Quibus jam excussis accessere demum alia quor[um] omnium synopsis sequenti pagina ponitur. Opus duobus voluminibus divisum Philippo IIII. regi catholico magno ....* Romae: Ex typographeio Iacobi Mascardi, 1649.

Dodoens 1554: Rembert Dodoens, *Cruijdeboeck: in den welcken die gheheele historie, dat es tgheslacht, tfatsoen, naem, natuere, cracht ende werckinghe, van den cruyden, niet alleen hier te lande wassende, maer oock van den anderen vremden*. Antuerpiæ: Jean van der Loe, 1554.

Engl. transl.: Henry Lyte, *A new herball, or, Historie of plants: wherein is contained the whole discourse and perfect description of all sorts of herbes and plants: their diuers and sundrie kindes: their names, natures, operations, & vertues: and that not onely of those which are heere growing in this our countrie of England, but of all others also of forraine realms commonly used in physicke, first set foorth in the douch or Almaigne toong*, Londinium: Ninian Newton, 1586.

Dodoens 1583: Rembert Dodoens, *Stirpium historiae pemptades sex sive libri xxx*. Antuerpiæ: Ex officina Christophori Plantini, 1583.

Duhamel du Monceau 1758: Henri Duhamel du Monceau, *La physique des arbres où il est traité de l'anatomie des plantes et de l'économie végétale*, 2 vols. Paris: H. L. Guérin et L. F. Delatour, 1758.

Goethe 1790: Johann Wolfgang von Goethe, *Versuch die metamorphose der pflanzen zu erklären.* Gotha: Carl Wilhelm Ettinger, 1790.

Jussieu 1718: Antoine de Jussieu, *Discours sur le progrès de la botanique.* Paris: Etienne Ganeau, 1718.

Jussieu 1789: Antoine-Laurent de Jussieu, *Genera plantarum secundum ordines naturales disposita, juxta methodum in horto regio parisiensi exaratam.* Paris: Herrisant et Barrois, 1789.

L'Écluse 1601: Charles de l'Écluse [Clusius], *Rariorum plantarum historia.* Antuerpiæ: Ex officina Christophori Plantini, 1601.

L'Obel and Pena 1571: Matthias de L'Obel and Pierre Pena, *Stirpium aduersaria noua, perfacilis vestigatio luculentaqne accessio ad priscorum praesertim Dioscoridis recentiorum materiam medicam quibus prope diem accedet altera pars qua coniectaneorum de plantis appendix, de succis medicatis et metallicis sectio antiquae et nouatae medicinae lectionum remedioru[m] thesaurus opulentissimus de succedantis libellus continentur authoribus Petro Pena & Mathia de Lobel medicis.* London: Thomae Purfoetii, 1571.

Laguna de Segovia 1543: Andrés Laguna de Segovia, *Aristotelis philosophorum principis de natura stirpium liber unus & alter.* J. Aquensis: Coloniae, 1543.

Lamarck 1783: Jean-Baptiste de Lamarck, *Encyclopédie méthodique ou par ordre de matières par une société de gens de lettres, de savants et d'artistes.* Tome 1: *Botanique.* Paris: Panckoucke, 1783.

Leoniceno 1492: Niccolò Leoniceno, *Plinij ac plurium aliorum auctorum qui de simplicibus medicaminibus scripserunt errores notati.* Ferrara: Laurentium de valentia et Andream de castronouo socios, 1492. Ed. and It. Transl.: Loris Premuda, *De Plinii in medicina erroribus.* Milan: Il Giardino di Esculapio, 1958.

Linnaeus 1736: Carl von Linné, *Fundamenta botanica quae majorum operum prodromi instar theoriam scientiae botanices per breves aphorismos tradunt.* Amsterdam: Salomon Schouten, 1736.

Linneaus 1737: Carl von Linné, *Genera plantarum eorumque characteres naturales secundum numerum, figuram, situm, et proportionem, omnium fructificationis partium.* Leiden: Conrad Wishoff, 1737.

Linneaus 1751: Carl von Linné, *Philosophia botanica in qua explicantur fundamenta botanica cum definitionibus partium, exemplis terminorum, observationibus rariorum, adjectis figuris aeneis.* Amsterdam and Stockholm: Kiesewetter, 1751.

Linneaus 1753: Carl von Linné, *Species plantarum exhibentes plantas rite cognitas ad genera relatas. Cum differentiis specificis, nominibus trivialibus, synonymis selectis, locis natalibus, secundum systema sexuale digestas*, 2 vols. Stockholm: Lars Salvius, 1753.

Magnol 1689: Pierre Magnol, *Prodromus historiæ generalis plantarum, in quo familiæ per tabulas disponuntur.* Monspelii: Gabriel & Honoré Pech, 1689.

Manardi 1529: Giovanni Manardi, *Epistolae medicinales in quibus multa recentiorum errata et antiquorum decretal reserentur.* Argentorati: Apud Joannem Schottum, 1529.

Mattioli 1544: Pietro Andrea Mattioli, *Di Pedacio Dioscoride Anazarbeo libri cinque della historia, et materia medicinale tradotti in lingua volgare italiana da M. Pietro Andrea Matthiolo Sanese Medico, con amplissimi discorsi, et comenti, et dottissime annotationi, et censure del medesimo interprete.* Venetiis: Niccolò de Bascarini, 1544.

Mattioli 1554: Pietro Andrea Mattioli, *Senensis commentarii, in libros sex Pedacii Dioscoridis Anazarbei, de materia medica, adjectis quam plurimis plantarum & animalium imaginibus.* Venetiis: in officina Erasmiana, apud Vincentium Valgrisium, 1554.

Melanchthon 1549: Philip Melanchthon, *Initia doctrinae physicae, dictata in Academia Vuitebergensi.* Vuitebergae: Iohannes Lufft, 1549 (ed. Karl Gottlieb Brettschneider, *Philippi Melanthonis opera quae supersunt omnia. Scripta ad historiam profanam et philosophiam spectantia* (Corpus reformatorum 13). Halle: Schwetschke, 1846, pp. 181–412).

Parker 1678: Samuel Parker, *Disputationes de Deo et providentia divina.* London: typis M. Clark, impensis Jo. Martyn, 1678.

Rabelais 1532–1534: Charles Esmangart and Eloi Johanneau (eds), *La vie de Gargantua et de Pantagruel* (*Œuvres de Rabelais*, vol 5). Paris: Dalibon, 1823.
Ray 1682: John Ray, *Methodus plantarum nova, brevitatis et perspicuitatis causa synoptice in tabulis exhibita; cum notis generum tum summorum tum sublaternorum characteristicis, observationibus nonnullis de siminibus plantarum et indice copioso*. Londini: Faithborne and Kersey, 1682.
Ray 1693: John Ray, *Historia plantarum generalis*. Londini: Samuel Smith and Benjamini Walford, 1693.
Ruel 1516: Jean Ruel, *Pedacii Dioscoridis Anazarbei de medicinali materia libri quinque. De virulentis animalibus, et venenis canerabioso, et corum notis, ac remediis libri quattuor.* Paris: Henrici Stephani, 1516.
Ruel 1536: Jean Ruel, *De natura stirpium libri tres*. Paris: Simon de Colines, 1536.
Ruel 1543: Jean Ruel, *De natura stirpium libri tres... cum indice omnium universi operis observatione dignorum copiosissimo*. Basilae: Frobeniana, 1543.
Scaliger 1556: Julius Caesar Scaliger, *In libros duos, qui inscribuntur De plantis, Aristotele autore, libri duo*. Lutetiae: Michael Vascosani, 1556.
Taurellus 1597: Nicolaus Taurellus, *Alpes caesae: hoc est, Andreae Caesalpini itali, monstrosa et superba dogmata, discussa et excussa ... Scriptum acre, acutum et limatum, philosophiae studiosis ad enodandas et enucleandas controuersias et disputationes physicas difficilimas admodum utile et necessarium*. Francofurti: apud M. Zachariam Palthenium, 1597.
Tournefort 1694: Joseph Pitton de Tournefort, *Éléments de botanique ou méthode pour connaître les plantes*, 3 vols. Paris: Imprimerie Royale, 1694.
Tournefort 1700–1703: Joseph Pitton de Tournefort, *Institutiones rei herbariæ*, 2 vol. Paris: Typographia Regia, 1700–1703.
Tyson 1699: Edward Tyson, *Orang-outang, sive homo sylvestris: or, The anatomy of a pygmie compared with that of a monkey, an ape, and a man*. London: T. Bennett and D. Brown, 1699.
Zalužanský 1592: Adam Zalužanský ze Zalužan [Adamus Mathiades Hradistenus], *Methodi herbariae libri tres*. Pragæ: G. Daczicenus, 1592.

Secondary literature

Amigues 2001: Suzanne Amigues, *Études de botanique antique*. Paris: De Boccard, 2001.
Andretta 2009: Elisa Andretta, "Mercati, Michele", in *Dizionario biografico degli Italiani*, vol. 73. Rome: Istituto dell'Enciclopedia Italiana, 2009, pp. 606–611.
Arber 1912: Agnes Arber, *Herbals, their origin and evolution, a chapter in the history of botany*. Cambridge: Cambridge University Press, 1912.
Arber 1950: Agnes Arber, *The natural philosophy of plant form*. Cambridge: Cambridge University Press, 1950.
Arcieri 1945: Giovanni P. Arcieri, *The circulation of the blood and Andrea Cesalpino of Arezzo*. New York: S. F. Vanni, 1945.
Atran 1990: Scott Atran, *Cognitive foundations of natural history. Towards an anthropology of science*. Cambridge: Cambridge University Press, 1990.
Baldassarri 2020: Fabrizio Baldassarri, "Early Modern Philosophy of Plants and the Unwelcome Guest: Pseudo-Aristotle's *De plantis*", in Maria Fernanda Ferrini and Guido Giglioni (eds), *Peri Phyton. Greek Botanical Treatises in the West and the East*. Macerata: EUM, 2020, pp. 237–264.
Baldassarri and Blank 2021: Fabrizio Baldassarri and Andreas Blank (eds), *Vegetative Powers: The Roots of Life in Ancient, Medieval, and Early Modern Natural Philosophy*. Cham: Springer, 2021.
Baldassarri 2023: Fabrizio Baldassarri, "A Renaissance Philosophy of Plants: Cesalpino's *De plantis* in Context", in Fabrizio Baldassarri and Craig Martin (eds), *Andrea Cesalpino: An Aristotelian Natural Philosopher in the Renaissance*. London: Bloomsbury, 2023 [forthcoming].

Barnes 2002: Jonathan Barnes, *Aristotle. Posterior Analytics*. Oxford: Clarendon Press, 2002 [1975].

Barnhart 1965: John Hendley Barnhart, *Biographical notes upon botanists*. Boston: G.K. Hall, 1965.

Bedini 2019: Gianni Bedini, *L'orto botanico di Pisa. The botanical Garden of Pisa*. Pisa: Pisa University Press, 2019.

Bellorini 2016: Cristina Bellorini, *The World of Plants in Renaissance Tuscany. Medicine and Botany*. Farnham and Burlington: Ashgate, 2016.

Beretta 2001: Ilva Beretta, "Illustration and Representation: Botany in the Renaissance", in Fabrizio Meroi and Claudio Pogliano (eds), *Immagini per conoscere: dal Rinascimento alla rivoluzione scientifica. Atti della giornata di studio, Firenze, Palazzo Strozzi, 29 ottobre 1999*. Florence: Leo S. Olschki, 2001, pp. 43–60.

Bernier 2013: Georges Bernier, *Darwin, un pionnier de la physiologie végétale: l'apport de son fils Francis*. Brussels: Académie royale de Belgique, 2013.

Blank 2023: Andreas Blank, "Cesalpino on Sensitive Powers and the Question of Divine Immanence" in Fabrizio Baldassarri and Craig Martin (eds), *Andrea Cesalpino: An Aristotelian Natural Philosopher in the Renaissance*. London: Bloomsbury, 2023 [forthcoming].

Böhner 1933–1936: Konrad Böhner, *Geschichte der Cecidologie: Ein Betrag zur Entwicklungsgeschichte naturwissenschaftlicher Forschung und ein Führer durch die Cecidologie der Alten*, 2 vols. Mittenwald: Nemayer, 1933–1936.

Bremekamp 1952: Cornelis Eliza Bertus Bremekamp, "A Re-examination of Cesalpino's Classification", *Acta botanica neerlandica* 1.4 (1952), pp. 580–593.

Bretin-Chabrol and Leduc 2009: Marine Bretin-Chabrol and Claudine Leduc, "La botanique antique et la problématique du genre", *Clio* 29 (2009), pp. 205–223.

Brisson 2023: Luc Brisson, "L'embryon est-il une plante? La réponse des néoplatoniciens", in Sarah Carvallo and Arnaud Macé (eds), *Analogies et modèles végétaux en médecine de l'Antiquité à la première Modernité*. Besançon: Presses Universitaires de Franche-Comté, 2023 [forthcoming].

Buoncristiani 2012: Anna Buoncristiani, "Andrea Cesalpino as a pioneer in classification", *Kos* 212 (2003), pp. 38–41.

Bylebyl 1972: Jerome Joseph Bylebyl, "Cesalpino and Harvey on the portal circulation", in Allen George Debus (ed.), *Science, medicine and society in the Renaissance: Essays to honor Walter Pagel*, vol. 2. New York: Science History Publications, 1972, pp. 39–52.

Callot 1971: Émile Callot, *La philosophie biologique de Goethe*. Paris: Rivière, 1971.

Calvi 1777: Giovanni Calvi, *Commentarium inservitурum historiae Pisani vireti botanici academici*. Pisa: De Pizzorni, 1777.

Capecci 1983: Angelo Capecci, "Finalismo e meccanicismo nelle ricerche biologiche di Cesalpino e Harvey", in Luigi Olivieri (ed.), *Aristotelismo veneto e scienza moderna*. Padua: Antenore, 1983, pp. 477–507.

Caruel 1858: Théodore Caruel, *Illustratio in hortum siccum Andreae Caesalpini*. Florence: Felice Le Monnier, 1858.

Cerrito 2021: Amalia Cerrito, "Disclosing the Hidden Life of Plants: Theories of the Vegetative Soul in Albert the Great's *De vegetabilibus et plantis*", in Fabrizio Baldassarri and Andreas Blank (eds), *Vegetative Powers*. Cham: Springer, 2021, pp. 105–122.

Chéreau 1873: Achille Chéreau, "André Cesalpin", in Amédée Dechambre (ed.), *Dictionnaire encyclopédique des sciences médicales*, vol. 14. Paris: Masson et Asselin, 1873, pp. 543–545.

CNRTL: Centre National de Ressources Textuelles et Lexicales (online ressource: https://www.cnrtl.fr/).

Colombero 1977: Carlo Colombero, "Il pensiero filosofico di Andrea Cesalpino", *Rivista critica di Storia della filosofia* 32.3 (1977), pp. 269–284.

Colombero 1980: Carlo Colombero, "Andrea Cesalpino e la polemica anti-aristotelica e anti-spinoziana", *Rivista critica di storia della filosofia* 35.4 (1980), pp. 343–356.

Condorelli 1975: Luigi Condorelli, "The discovery of phyllotaxy by Andrea Cesalpino", *Minerva medica* 66.83 (1975), pp. 4460–4463.

Daugey 2015: Fleur Daugey, *Les plantes ont-elles un sexe? Histoire d'une découverte*. Paris: Ulmer, 2015.

Davy de Virville 1954: Adrien Davy de Virville (ed.), *Histoire de la botanique en France*. Paris: Société d'édition d'enseignement supérieur, 1954.

Davy de Virville and Leroy 1969: Adrien Davy de Virville and Jean-Fraçois Leroy, "Botanique à la Renaissance", in Edmond Bauer and René Taton (eds), *La science moderne de 1450 à 1800*. Paris: Presses Universitaires de France, 1969, pp. 177–188.

Delaporte 2011: François Delaporte, *Le second règne de la nature: essai sur les questions de végétalité au XVIIIe siècle*. Paris: Archives contemporaines, 2011 [first edition 1979] (Engl. transl. Arthur Goldhammer, *Nature's Second Kingdom: Explorations of Vegetality in the Eighteenth Century*. Cambridge, MA, and London: MIT Press, 1982).

Del Soldato 2020: Eva Del Soldato, *Early Modern Aristotle: On the Making and Unmaking of Authority*. Philadelphia: University of Pennsylvania Press, 2020.

Del Soldato 2023: Eva Del Soldato, "Plato and Cesalpino's Aristotelianism: A Revealing Marginality" in Fabrizio Baldassarri and Craig Martin (eds), *Andrea Cesalpino: An Aristotelian Natural Philosopher in the Renaissance*. London: Bloomsbury, 2023 [forthcoming].

Delcomminette 2018: Sylvain Delcomminette, *Aristote et la nécessité*. Paris: Vrin, 2018.

De Vos 2010: Paula De Vos, "European materia medica in historical texts: longevity of a tradition and implications for future use", *Journal of Ethnopharmacology* 132.1 (2010), pp. 28–47.

De Wit 1992: Hendrik Cornelis Dirk de Wit: "Césalpin et la physiologie des plantes et des humains: la circulation du sang", in Hendrik Cornelis Dirk de Wit (ed.), *Histoire du développement de la biologie*, vol. 1. Lausanne: Presses polytechniques et universitaires romandes, 1992, pp. 350–353.

Dughi 1957: Raymond Dughi, "Tournefort dans l'histoire de la botanique", in Roger Heim (ed.), *Tournefort*. Paris: Muséum National d'Histoire Naturelle, 1957, pp. 131–186.

Edwards 1976: William Edwards, "Niccolò Leoniceno and the origins of humanist discussion of method", in Edward Patrick Mahoney (ed.), *Philosophy and humanism. Renaissance essays in honor of Paul Oskar Kristeller*. Leiden: Brill, 1976.

Egmond et al. 2007: Florike Egmond, Paul Hoftijzer, Robert Visser (eds), *Carolus Clusius. Towards a cultural history of a Renaissance naturalist*. Amsterdam: Koninklijke Nederlandse Akademie van Wetenschappen, 2007.

Egmond 2010: Florike Egmond, *The World of Carolus Clusius. Natural history in the making 1550–1610*. London: Pickering and Chatto, 2010.

Fatigati 2021: Michael Fatigati, "Avicenna on Vegetative Faculties and the Life of Plants", in Fabrizio Baldassarri & Andreas Blank (eds), *Vegetative Powers: The Roots of Life in Ancient, Medieval and Early Modern Natural Philosophy*. Cham: Springer. 2021, pp. 73–89.

Ferri 1997. Sara Ferri, *Pietro Andrea Mattioli (Siena 1501 – Trento 1578). La vita le opere con l'identificazione delle piante*. Ponte San Giovanni: Quattroemme, 1997.

Fornasier 2020: Matteo Fornasier, "Teleology in Renaissance Science", in Marco Sgarbi (ed.), *Encyclopedia of Renaissance Philosophy*. Cham: Springer, 2020, https://doi.org/10.1007/978-3-319-02848-4_408-1.

Findlen 1999: Paula Findlen, "The Formation of a Scientific Community: Natural History in Sixteenth-Century Italy", in Anthony Grafton and Nancy Siraisi (eds), *Natural Particulars, Nature and the Disciplines in Renaissance Europe*. Cambridge, MA, and London: MIT Press, 1999, pp. 369–400.

Fiorentino 1911: Francesco Fiorentino, "Vita ed opere di A. Cesalpino", in *Studi e retratti della Rinascenza*. Bari: Laterza, 1911.

Franconie 1997: Hélène Franconie, "Maïs et millets dans les dialectes européens", *Journal d'agriculture traditionnelle et de botanique appliquée* 39 (1997), pp. 149–178.

Ghellinck 1942: Joseph de Ghellinck, "L'entrée d'*essentia*, *substantia* et autres mots apparentés dans le latin medieval", *Archivium latinitatis medii aevi* 16 (1942), pp. 77–112.

Givens et al. 2006: Jean Ann Givens, Karen Meier Reeds and Alain Touwaide (eds), *Visualizing Medieval Medicine, 1200–1500*. Aldershot: Ashgate, 2006.

Greene 1909: Edward Lee Greene, *Landmarks of Botanical History. Part I.* Washington: Smithsonian Institution, 1909.

Greene 1983: Edward Lee Greene and Frank Egerton (ed.), *Landmarks of Botanical History. Part II.* 2 vol. Stanford: Stanford University Press, 1983.

Hallé et al. 1978: Francis Hallé, Roelof Arend Albert Oldeman and Philip Barry Tomlinson, *Tropical trees and forests. An architectural analysis.* Berlin, Heidelberg, New York: Springer, 1978.

Hallé 1999: Francis Hallé, *Éloge de la plante*. Paris: Seuil, 1999.

Harvey-Gibson 1919: Robert John Harvey Gibson, *Outlines of the history of botany*. London: A. & C. Black, 1919 (ed. New York: Arno Press, 1981).

Heideklang 2023: Julia Heideklang, "Paratextual Debates in *De plantis* (1583): On the Best Form of Botanical Prose, Garden and Things, and the Author-Figure of Cesalpino", in Fabrizio Baldassarri and Craig Martin (eds), *Andrea Cesalpino: An Aristotelian Natural Philosopher in the Renaissance.* London: Bloomsbury, 2023 [forthcoming].

Heim 1957: Roger Heim (ed.), *Tournefort*. Paris: Muséum National d'Histoire Naturelle, 1957.

Helbing 2008: Mario Otto Helbing, "Mechanics and natural philosophy in late 16th-century Pisa: Cesalpino and Buonamici, humanist masters of the faculty of arts", *Boston studies in the philosophy of science* 254 (2008), pp. 185–193.

Henning 2009: Boris Henning, "The Four Causes", *Journal of Philosophy* 106.3 (2009), pp. 137–160.

Hiernaux and Tresnie 2023a: Quentin Hiernaux and Corentin Tresnie, "Aristotelian Metaphysics of the Vegetative Soul and Early Modern Plant Physiology: Comparison between Plant Functions in Aristotle, Pseudo-Aristotle, and Cesalpino", in Fabrizio Baldassarri and Craig Martin (eds), *Andrea Cesalpino: An Aristotelian Natural Philosopher in the Renaissance.* London: Bloomsbury, 2023 [forthcoming].

Hiernaux and Tresnie 2023b: Quentin Hiernaux and Corentin Tresnie, "L'analogie dans la botanique d'Andrea Cesalpino", in Sarah Carvallo and Arnaud Macé (eds), *Analogies végétales en médecine de l'Antiquité à l'Âge classique.* Besançon: Presses Universitaires de Franche-comté, 2023 [forthcoming].

Hoefer 1872: Ferdinand Hoefer, *Histoire de la botanique, de la minéralogie et de la géologie depuis les temps les plus reculés jusqu'à nos jours*. Paris: Hachette, 1872.

Hoquet 2005: Thierry Hoquet (ed.), *Les fondements de la botanique. Linné et la classification des plantes*. Paris: Vuibert, 2005.

Ingegno 1988: Alfonso Ingegno, "The New Philosophy of Nature", in Charles Bernard Schmitt and Quentin Skinner (eds), *The Cambridge History of Renaissance Philosohy.* Cambridge: Cambridge University Press, 1988, pp. 236–263.

Jensen 2001: Kristian Jensen, "Description, Division, Definition – Caesalpinus and the Study of Plants as an Independent Discipline", in Marianne Pade (ed.), *Renaissance Reading of the Corpus aristotelicum*. Copenhagen: Museum Tusculanum Press, 2001, pp. 185–206.

Kusukawa 2012: Sachiko Kusukawa, *Picturing the Book of Nature. Image, Text, and Argument in Sixteenth-Century Human Anatomy and Medical Botany*. London and Chigaco: University of Chicago Press, 2012.

Larson 1971: James L. Larson, *Reason and experience: the representation of natural order in the work of Carl von Linné.* Berkeley: University of California Press, 1971.

Lefèvre 2018: Denis Lefèvre, *Des racines et des gènes. Une histoire mondiale de l'agriculture*, 2 vol. Paris: rue de l'échiquier, 2018.

Leroy 1957: Jean F. Leroy, "Tournefort et la classification végétale", in Roger Heim (ed.), *Tournefort.* Paris: Muséum National d'Histoire Naturelle, 1957, pp. 187–206.

Lovejoy 1936: Arthur Lovejoy, *The great chain of beings: A study of the history of an idea.* Cambrige, MA: Harvard University Press, 1936.

Mägdefrau 1978: Karl Mägdefrau, "Andrea Cesalpino (Andrea Caesalpinus) (1519–1603)", in Charles Coulston Gillispie (ed.), *Dictionary of scientific biography.* New York: Scribner's, 1978, vol. 15, pp. 80–81.

Magnin-Gonze 2009: Joëlle Magnin-Gonze, *Histoire de la botanique*. Paris: Delachaux et Niestlé, 2009.

Mayerhöfer 1959: Josef Mayerhöfer, *Lexikon der Geschichte der Naturwissenschaften Biographien, Sächwörter und Bibliographien. Mit einer Einführung "Die Zeitalter der Naturforschun" und einer Übersichtastabelle.* Vienna: Hollinek, 1959.

Mayr 1982: Ernst Mayr, *The growth of biological thought. Diversity, Evolution, and Inheritance*. Cambridge and London: The Belknap Press of Harvard University Press, 1982.

Meyer 1854–1857: Ernst Heinrich Friedrich Meyer, *Geschichte der Botanik*, 4 vols. Königsberg: Bornträger, 1854–1857.

Mix 2018: Lucas John Mix, *Life Concepts from Aristotle to Darwin: On Vegetable Souls.* SpringerLink : Online service, 2018.

Moggi 2009: Guido Moggi, "L'erbario di Andrea Cesalpino. The Andrea Cesalpino Herbarium", in Mauro Raffaelli (ed.), *Il museo di storia naturale dell'università degli studi di Firenze: le collezioni botaniche / The Museum of Natural History of the University of Florence: The Botanical Collections*. Florence: Firenze university Press, 2009, pp. 1–20.

Moreau 2023: Elizabeth Moreau, "The Pharmacological Theory of Andrea Cesalpino (1593)", in Fabrizio Baldassarri and Craig Martin (eds), *Andrea Cesalpino: An Aristotelian Natural Philosopher in the Renaissance.* London: Bloomsbury, 2023 [forthcoming].

Morton 1981a: Alan G. Morton, *History of Botanical Science: An Account of the Development of Botany from Ancient Times to the Present Day.* London: Academic Press Inc, 1981.

Morton 1981b: Alan G. Morton, "Marginalia to Andrea Cesalpino's work on botany", *Archives of Natural History* 10.1 (1981), pp. 31–36.

Morton 1987: Alan G. Morton, "A letter of Andrea Cesalpino", *Archives of Natural History* 14.2 (1987), pp. 169–73.

Mugnai Carrara 1994: Daniela Mugnai Carrara, "Niccolò Leoniceno e Giovanni Mainardi: aspetti epistemologici dell'Umanesimo medico", in Marco Bertozzi (ed.), *Alla corte degli estensi. Filosofia, arte et cultura a Ferrara nei secoli XV e XVI. Atti del convegno internazionale di studi, Ferrara, 5–7 marzo, 1992.* Ferrara: Universita degli studi, 1994.

Müller-Wille 2005: Staffan Müller-Wille, "La science de Bacon en action: la place de Linné dans l'histoire de la taxonomie", in Thierry Hoquet (ed.), *Les fondements de la botanique. Linné et la classification des plantes.* Paris: Vuibert, 2005.

Negbi 1995: Moshe Negbi, "Male and Female in Theophrastus's Botanical Works", *Journal of the history of biology* 28 (1995), pp. 317–332.

Nepi 2007: Chiara Nepi, "La 'slegatura' dell'erbario di Andrea Cesalpino (1525–1603)", *Museologia scientifica* 1 (2007), pp. 50–54.

Nepi and Gusmeroli 2008: Chiara Nepi and Enrico Gusmeroli (eds), *Gli erbari aretini da Andrea Cesalpino ai giorni nostri*. Florence: Firenze University Press, 2008.

Nisard 1864: Jean Marie Napoléon Désiré vard (ed.), *Les agronomes latins: Caton, Varron, Columelle, Palladius*. Paris: Firmin Didot Frères, fils et Cie, 1864.

Ogilvie 2006: Brian Ogilvie, *The Science of Describing*. Chicago: University of Chicago Press, 2006.

Olariu 2018: Dominic Olariu, "'Kräuterautopsie' im Jahr 1487", *Geschichte der Pharmazie* 70.3 (2018), pp. 29–51.

O'Malley and Meyers 2008: Therese O'Malley and Amy Meyers (eds), *The Art and History of Botanical Painting and Natural History Treatises*. Washington, DC: The National Gallery of Art, 2008.

Pagel 1953: Walter Pagel, "Die Stellung Caesalpins und Harveys in der Entdeckung und ideologie des Blutkreislaufs", *Sudhoffs Archiv für Geschichte der Medizin und der Naturwissenschaften* 37 (1953), pp. 319–328.

Pagel 1975: Walter Pagel, "Cesalpino's Peripatetic Questions", *History of Science* 13 (1975), pp. 130–138.

Panarelli 2020: Marilena Panarelli, "Albert the Great's *De vegetabilibus* and its unique position among the medieval commentaries on *De plantis*" in Maria Fernanda Ferrini and Guido Giglioni (eds), *Peri Phyton.*

Greek Botanical Treatises in the West and the East. Macerata: edizioni Università di Macerata, 2020, pp. 137–162.

Panarelli 2021: Marilena Panarelli, "Can Plants Desire? Aspects of the Debate on desiderium naturale", in Fabrizio Baldassarri and Andreas Blank (eds), *Vegetative Powers*. Cham: Springer, 2021, pp. 91–104.

Partington 1962: James Riddick Partington, *A history of chemistry*. London: Macmillan, 1962.

Peller 1949: Sigismund Peller, "Harvey's and Cesalpino's role in the history of medicine", *Bulletin of the History of Medicine* 23.3 (1949), pp. 213–235.

Pieragnoli 1853: Misael Pieragnoli, *Della vita e delle opere di Michele Mercati juniore*. San Miniato: Stamperia Ristore, 1853.

Plantefol and Prévot 1962: Lucien Plantefol and Anne-Marie Prévost, "La notion d'étamine à travers diverses dénominations", *Revue philosophique de la France et de l'étranger* 152 (1962), pp. 145–172.

Plantefol 1969: Lucien Plantefol, "Duhamel Dumonceau", *Dix-huitième siècle* 1 (1969), pp. 123–137.

Premuda 1987–1988: Loris Premuda, "Empirismo e scienza in Prospero Alpini (1553–1616)", *Acta medicae historiae patavina* 34 (1987–1988), pp. 71–81.

Ramminger 2005: Johann Ramminger, "A commentary? Ermolao Barbaro's Supplement to Dioscorides", in Marianne Pade (ed.), *Noctes Neolatinae. Neo Latin text and studies*, vol. 4 *On Renaissance commentaries*. Georg Olms Verlag: Hildesheim, Zürich, New York, 2005, pp. 65–85.

Raven 1950: Charles E. Raven, *John Ray naturalist. His life and works*. Cambridge: Cambridge University Press, 1950.

Reeds 1991: Karen Meier Reeds, *Botany in medieval and Renaissance universities*. London and New York: Garland Publishing, 1991.

Repici 2005: Luciana Repici, "Andrea Cesalpino e la botanica antica", *Rinascimento* 45 (2005), pp. 47–87.

Richelet 1759: Pierre Richelet, *Dictionnaire de la langue française, ancienne et modern*, 3 vols. Lyon: Frères Duplain, 1759.

Rothschuh 1973: Karl E. Rothschuh, *History of Physiology*. New York: Robert E. Krieger Publishing Company, 1973.

Sabbatani 1921: Luigi Sabbatani, *Lucae Ghini morbi gallici curandi ratio perbrevis: la cura del morbo gallico nelle lezioni di Luca Ghini*. Venice: Ferrari, 1921.

Sachs 1875: Julius von Sachs, *Geschichte der Botanik vom 16. Jahrhundert bis 1860*. Munich: Didenbourg, 1875 (Engl. transl. Henry Edward Fowler Garnsey, *History of botany (1530–1560)*. Oxford: Clarendon Press, 1890) (Fr. transl. Henri de Varigny, *Histoire de la botanique du XVIe siècle à 1860*. Paris: Reinwald et Cie, 1892).

Saint-Lager 1884: Jean Baptiste Saint-Lager, "Remarques historiques sur les mots plantes mâles et plantes femelles", *Annales de la société botanique de Lyon* 11 (1883–1884), pp. 1–48.

Saint-Lager 1885: Jean Baptiste Saint-Lager, "Histoire des herbiers", *Publications de la société linnéenne de Lyon* 13 (1885–1886), pp. 1–120.

Sauvé Meyer 1992: Susan Sauvé Meyer, "Aristotle, Teleology, and Reduction", *Philosophical Review* 101 (1992), pp. 791–825.

Schmitt 1972: Charles B. Schmitt, "The faculty of Arts at Pisa at the time of Galileo", *Physis* 14 (1972), pp. 243–272.

Sgarbi 2023: Marco Sgarbi, "Andrea Cesalpino's Epistemology: A Method for Botany", in Fabrizio Baldassarri and Craig Martin (eds), *Andrea Cesalpino: An Aristotelian Natural Philosopher in the Renaissance*. London: Bloomsbury, 2023 [forthcoming].

Singer 1950: Charles Singer, *A history of biology. Revised edition. A general introduction to the study of living things*. New York: Henry Schuman, 1950.

Sprague 1939: Thomas Archibald Sprague, "The herbal of Valerius Cordus", *Botanical Journal of the Linnean Society* 52.341(1939), pp. 1–113.

Sprague 1950: Thomas Archibald Sprague, "The evolution of botanical taxonomy from Theophrastus to Linnaeus", in Linnean Society of London, *Lectures on the Development of Taxonomy Delivered in the Rooms of the Linnean Society During the Session 1948–49*. London: Linnean Society, 1950, pp. 1–23.

Stafleu 1971: Frans Antonie Stafleu, *Linnaeus and the Linnaeans : the spreading of their ideas in systematic botany : 1735–1789*. Utrecht: Oosthoek's Uitgeversmaatschappij, 1971.

Stafleu and Cowan 1976–1988: Frans Antonie Stafleu and Richard Sumner Cowan, *Taxonomic Literature: A selective guide to botanical publications and collections with dates, commentaries and types*, 7 vols. (Regnum vegetabile 94, 98, 105, 110, 112, 115, 116). Utrecht: Bohn, Scheltema and Holkema, 1976–1988.

Stearn 1966: William Thomas Stearn, *Botanical latin*. New York: Hafner publishing Co., 1966.

Stefanaki et al. 2019: Anastasia Stefanaki, Henk Porck, Ilaria Maria Grimaldi, Nikolaus Thurn, Valentina Pugliano, Adriaan Kardinaal, Jochem Salemink, Gerard Thijsse, Claudine Chavannes-Mazel, Erik Kwakkel, Tinde van Andel, "Breaking the silence of the 500-year-old smiling garden of everlasting flowers: The En Tibi book herbarium", *PLoS ONE* 14.6 (2019), e0217779.

Sprengel 1807–1808: Kurt Sprengel, *Hisotria rei herbariae*, 2 vol. Amstelodami: Sumtibus tabernae librariae et artum, 1807–1808.

Sprengels 1817: Kurt Sprengels, *Geschichte der Botanik*. Altenburg and Leipzig: Brockhaus, 1817.

Stevens 1994: Peter F. Stevens, *The development of biological systematics: Antoine-Laurent de Jussieu, Nature, and the Natural System*. New York: Columbia University Press, 1994.

Stevenson 1947: Ian Pretyman Stevenson, "John Ray and his contributions to plant and animal classification", *Journal of the History of Medicine and Allied Sciences* 2 (1947), pp. 250–261.

Strazzoni 2016: Andrea Strazzoni, "Cesalpino, Andrea", in Marco Sgarbi (ed.), *Encyclopedia of Renaissance Philosophy*. Cham: Springer International Publishing, 2016, pp. 1–3.

Swan 2002: Claudia Swan, "From Blowfish to Flower Still Life Painting: Classification and its Images, ca. 1600", in Pamela H. Smith and Paula Findlen (eds), *Merchants and Marvels: Commerce, Art, and the Representation of Nature in Early Modern Europe*. New York: Routledge, 2002, pp. 109–136.

Swan 2006: Claudia Swan, "The Uses of Realism in Early Modern Illustrated Botany", in Jean Ann Givens, Karen Meier Reeds and Alain Touwaide (eds), *Visualizing Medieval Medicine, 1200–1500*. Aldershot: Ashgate, 2006, pp. 239–250.

Swan 2011: Claudia Swan, "Illustrated Natural History", in Susan Dackermann (ed.), *Prints and the Pursuit of Knowledge*. Cambridge, MA: Harvard University Art Museums, 2011, pp. 186–191.

Thiers 2020: Barbara M. Thiers, *Herbarium. The quest to preserve and classify the world's plants*. Portland: Timber Press, 2020.

Thorndike 1966: Lynn Thorndike, "Cesalpino's vue of Nature", in Lynn Thorndike (ed.), *A History of Magic and Experimental Science*, vol. 6. New York: Columbia University Press, 1966, pp. 325–338.

Toni 1907: Giovanni Baptista de Toni, "I placiti di Luca Ghini intorno a piante descritte nei commentarii al Dioscoride di P.A. Mattioli", *Memorie del reale istituto veneto di scienze, lettere ed arti* 27.8 (1907), pp. 1–49.

Touwaide 1984: Alain Touwaide, *Le Traité de matière médicale de Dioscoride à la Renaissance*. Louvain-la-Neuve: Presses de l'Université, 1984.

Touwaide 2000: Alain Touwaide, "*Loquantur ipsi ut velint … dumtaxat non ignorent quis serpens sit tiro* – Leoniceno's contribution to the construction of Renaissance botanical lexicon", in Albert Derolez, Wouter Bracke and Herwig Deumens (eds), *Medical Latin from the Middle Ages to the Early Modern Times. European Science Foundation. Proceedings of the Symposium held in Brussels, September 1999*. Leuven: Peeters-Orientaliste, 2000, pp. 151–173.

Touwaide 2007: Alain Touwaide, "Leoniceno, Niccolò", in Noretta Koertge (ed.), *New Dictionary of Scientific Biography*, vol 4. Farmington Hills, MI: Charles Scribner's Sons, 2007, pp. 264–267.

Touwaide 2016: Alain Touwaide, *A Census of Greek Medical Manuscripts. From Byzantium to the Renaissance* (Medicine in the Medieval Mediterranean 6). London: Routledge, 2016.

Touwaide 2017: Alain Touwaide, “Renaissance Herbalism. The Perils of Humanism”, *Mefisto. Rivista di medicina, filosofia, storia* 2 (2017), pp. 1–18.
Trégo 2012: Kristell Trégo, “Substance, sujet, acte. La première réception latine d’Aristote: Marius Victorinus et Boèce”, *Études philosophiques* 101 (2012), pp. 233–256.
Trewavas 2014: Anthony Trewavas, *Plant behaviour and Intelligence*. Oxford: Oxford University Press, 2014.
Ventura 2013: Iolanda Ventura, “Changing Representations of Botany in Encyclopaedias from the Middle Ages to the Renaissance”, in Anja-Silvia Goeing, Anthony Thomas Grafton, and Paul Michel (eds), *Collectors’ Knowledge: What Is Kept, What Is Discarded*. Leiden and Boston: Brill, 2013, pp. 97–144.
Viviani 1922: Ugo Viviani, *Vita ed opere di Andrea Cesalpino*. Arezzo: Ugo Viviani, 1922.
Wear 1990: Andrew Wear, “The heart and blood from Vesalius to Harvey” in R. C. Olby, G. N. Cantor, J. R. R. Christie & M. J. S. Hodge (eds), *Companion to the History of Modern Science*. London: Routledge, 1990, pp. 571–574.
Zirkle 1936: Conrad Zirkle, “Animal impregnated by the wind”, *Isis* 69.25 (1936), pp. 95–130.

Index of Plant Names in *De plantis*, Book I

This index combines the index included by Cesalpino in *De plantis*, and the inventory of his herbarium of dried plants conserved at the Museo di Botanica in Florence. The herbarium contains many species mentioned in *De plantis* and identified by Théodore Caruel in 1853. Names followed by a question mark are not listed by Caruel, and their identification, based solely on the text of *De plantis*, is only probable or uncertain.[604] When the index of *De plantis* contains very similar names, we added these names in square brackets as well as their description in other books. In these cases, the common English names are also hypothetical.

Greek names are mostly taken from Caruel.[605] When his identification mentions Theophrastus (Th.) or Dioscorides (Di.), their initials are given. The same applies to Latin names found in Pliny the Elder (Pl.). Several Greek names, not found in Caruel, are from Theophrastus' indices of Greek names, and of scientific names of plants.[606] However, to avoid confusion, when a Greek name equivalent to the Latin name is attested (for example in Theophrastus), but refers to a species different from that mentioned by Cesalpino, the Greek name is not used.

Latin binomial names are the official 2022 species names. These can therefore differ from the basionyms (for example Linnaean), or synonyms listed by Caruel (1853).

Current common names were identified with the help of www.tela-botanica.org and Wikipedia, and may not be exhaustive. Synonyms are separated by a comma; names of different species, where there are several, are separated by a semicolon.

References to occurrences from Book I should be read as name of chapter followed by name of paragraph in the translated version, for example: chapter 1, second paragraph (1: §2).

References to specific descriptions of a species in the other books of *De plantis* are provided; simple mentions of the name of a species elsewhere in *De plantis* are not indexed. Books are identified in Roman numerals, followed by the chapter number in Arabic numerals.

A term followed by "(genus)" indicates that this term does not refer to a particular species, or that it is not possible to identify exactly which species is mentioned, or that Cesalpino is referring to the name of a genus rather than to the name of a species. Common names are therefore given in the plural. Cesalpino does not always specify the epithet of the species that he is describing, but context allows us to identify it in most cases. For example, "*Buglossum*" is most likely a reference to the species that he calls "*Buglossa hispana*", because it is the only species of bugloss present in his herbarium. These more precise identifications, as well as variations in spelling or possible synonyms, are included in brackets after the Latin name mentioned in Book I. When the name of the genus and the context do not allow us to identify the species collected by Cesalpino or described elsewhere in *De plantis*, the possible names of species are listed and separated by a semicolon.

The index both refers to species as plants in their entirety, and to specific parts of species bearing the same name in Latin: for example, the apple or the apple tree (*Malus*), the rose or the rose bush (*Rosa*).

---

604 When they exist in Cesalpino's own index, these plants are described in other books from *De plantis*. Deeper botanical study of these descriptions would be necessary to identify the species.

605 Caruel 1853.

606 Amigues 2006. (Index available online at https://uses.plantnet project.org/fr/Noms_grecs_de_Th%C3%A9ophraste and at https://uses.plantnet project.org/fr/Noms_scientifiques_des_plantes_de_Th%C3%A9ophraste).

Abies {anc.gr. Ἐλάτη; curr.lat. *Abies alba* Mill.} ■ (European) silver fir: 1: §10; 2: §17; 4: §46; 10: §93 ▸ III, 53
Acanaceum (genus) ■ Acanaceous, Thornies: 1: §6; 7: §74; 10: §108; 11: §115; 12: §120; 14: §151
Aconitum (genus) {anc.gr. Ἀκόνιτον; curr.lat. *Aconitum (*genus*); Aconitum lycoctonum* L.?;} ■ Aconitum; Wolf's-bane, Northern wolf's-bane: 10: §104 ▸ XV, 32, 33, 34
Aculeatus (genus) ■ Thornies: 12: §128
Adiantum (Adiantum album) {curr.lat. *Adiantum capillus-veneris* L.} ■ Southern maidenhair fern, Black maidenhair fern, Maidenhair fern, Venus hair fern: 2: §21; 14: §148 ▸ XVI, 12
Adiantum (Adiantum nigrum, Pl. Ruta muraria) {anc.gr. Ἀδίαντον (Th. Σαξίφραγον); curr.lat. *Asplenium ruta-muraria* L.} ■ Wall-rue: 2: §21; 14: §148 ▸ XVI, 12–13
Agrifolius (Agrifoglo) {curr.lat. *Ilex aquifolium* L.?} ■ Holly, Common holly, English holly, European holly: 11: §115 ▸ III, 6
Allium {curr.lat. *Allium scorodoprasum* L.} ■ Sand leek, Rocambole: 4: §41; 5: §52 ▸ X, 15
Amentaceum ■ Amentaceous: 7: §73
Amentiferum (genus); (Amentum) {curr.lat. *Amentiferae; Hamamelididae (subclass)*} ■ Amentiferous (catkin, ament): 7: §72
Amigdala {curr.lat. *Prunus dulcis* (Mill.) D.A.Webb?} ■ Almond (tree): 6: §60, §67; 8: §80, §82; 9: §91; 10: §111; 14: §151
Anchusa {anc.gr. Ἄγχουσα; curr.lat. *Echium italicum* L.?; *Echium plantagineum* L.?; *Buglossoides arvensis* (L.) I.M.Johnst.?} ■ Italian viper's bugloss (Lady Campbell weed, Pale bugloss); Purple viper's-bugloss (Paterson's curse); Field gromwell (Corn gromwell, Bastard alkanet): 7: §70 ▸ XI, 8, 9
Aphaca (genus) {anc.gr. Ἀφάκη; curr.lat. *Taraxacum (genus)?*} ■ Dandelions: 12: §132 ▸ VI, 12; XIII,3
Apium (genus) {anc.gr. σέλινον; curr.lat. *Apium (genus); Apium graveolens* L.?} ■ Marshworts; Celeri: 12: §125, §132 ▸ VIII, 38
Arbor (genus) {anc.gr. Théo. Δένδρον; } ■ Trees: 1: §5, §12; 2: §15; 3: §30–32, §35–37; 4: §38, §42, §46; 5: §49, §51; 8: §82; 9: §88, §91; 10: §93, §100, §110–§112; 11: §114, §116; 12: §118, §119, §124, §128, §133; 13: §137, §138, §140, §144; 14: §153, §157; ▸ II, III
Arbutus {curr.lat. *Arbutus unedo* L.?} ■ Strawberry tree: 3: §31 ▸ III, 19
Aristolochia longa {anc.gr. Ἀριστολοχία μάκρα; curr.lat. *Aristolochia clematitis* L.} ■ European birthwort: 12: §131 ▸ XV, 5
Aristolochia rotunda {anc.gr. Ἀριστολοχία στρογγύλη; curr.lat. *Aristolochia rotunda* L.} ■ Smearwort; Round-leaved birthwort: 1: §6; 12: §131 ▸ XVI, 5
Aron {anc.gr. Ἄρον; curr.lat. *Arum italicum* Mill.} ■ Italian arum, Italian lords-and-ladies: 12: §131 ▸ V, 36
Asclepias {anc.gr. Ἀσκληπιάς; curr.lat. *Vincetoxicum hirundinaria* Medik.} ■ White swallow-wort: 10: §98 ▸ VI, 72
Asparagus (genus) {anc.gr. Ἀσπάραγος ἄγριος; curr.lat. *Asparagus (genus); Asparagus tenuifolius* Lam.?; *Asparagus acutifolius* L.?} ■ Asparagus; *Asparagus tenuifolius*; Wild asparagus: 11: §115; 12: §127 ▸ V, 26
Asphodolum (Genus)(Asphodelus) {anc.gr. Ἀσφοδελός; curr.lat. *Asphodelus (genus);*} ■ Asphodels: 12: §132 ▸ X, 26, 29, 30
Aurantia {anc.gr. Χρυσόμηλον; curr.lat. *Citrus cinensis* (L.) Osbeck?} ■ Orange tree; Orange: 10: §98 ▸ III, 59
(Avellana) fraxinus (Melia) {anc.gr. Μελία; curr.lat. *Fraxinus ornus* L.} ■ Manna ash, South European flowering ash: 10: §93 ▸ II, 13
Brassica {curr.lat. *Brassica oleracea* L.?} ■ Cabbage: 12: §127 ▸ VIII, 54
Buglossum (Buglossa hispana) {anc.gr. Λύκοψις; curr.lat. *Pentaglottis sempervirens* (L.) Tausch ex L.H.Bailey} ■ Evergreen bugloss, Green alkanet, Alkanet: 10: §96 ▸ XI, 4
Bulbaceum (genus) ■ Bulbaceous: 4: §41, §42; 14: §151 ▸ X, 1
Caepa (Bulbus[agrestis]) {anc.gr. Th. Κρόμμυον; curr.lat. *Allium cepa* L.?} ■ Onion: 10: §110; 12: §127, §132 ▸ X, 19
Calamentus {anc.gr. Καλαμίνθη; curr.lat. *Nepeta cataria* L.} ■ Catnip, Catswort, Catwort, Catmint: 7: §74 ▸ XI, 63

Calamus (genus?) {curr.lat. *Arundo donax* L. ?; (genus: *Juncus*?)} ■ Giant cane, Elephant grass, Carrizo, Spanish cane, Colorado river reed, Wild cane, Giant reed; (Rushes): 3: §35; 4: §43 ‣ IV, 59

Cannabis {anc.gr. Κάνναβις; curr.lat. *Cannabis sativa* L.} ■ Hemp: 7: §73 ‣ IV, 16

Carduus (Carduus sylvestris) {anc.gr. Σκόλυμος; curr.lat. *Cynara cardunculus* L.?; *Onopordium acanthium* L.?; *Carduus nutans* L.?} ■ Globe artichoke; Cardoon; Artichoke thistle; Thistle: 8: §84 ‣ XIII, 40, 44

Castanea {curr.lat. *Castanea sativa* Mill.?} ■ Sweet chestnut, Spanish chestnut, Chestnut: 7: §72; 8: §77, §79, §82 ‣ II, 4

Catanance (Catanance altera) {anc.gr. Ἡλιοτρόπιον μέγα; curr.lat. *Impatiens noli-tangere* L.} ■ Touch-me-not balsam: 10: §101 ‣ VI, 68

Cerasus {curr.lat. *Prunus cerasus* L.?} ■ Cherry tree; Sour cherry, Tart cherry, Dwarf cherry: 9: §87, §90 ‣ II, 20

Cham[a]edaphne {anc.gr. Χαμαιδάφνη; curr.lat. *Ruscus Hypophyllum* L.} ■ Dutch Butcher's Broom: 10: §110 ‣ V, 30; VIII, 29

Chamaeleon {anc.gr. Χαμαιλέων; curr.lat. *Carlina acaulis* L.} ■ Stemless carline thistle, dwarf carline thistle, Silver thistle: 10: §110 ‣ XIII, 38, 39

Chamaemelon (Chamaemelum; Anthemis) {anc.gr. Χαμαίμηλον; Ἀνθεμίς; curr.lat. *Matricaria Chamomilla* L.} ■ Chamomile, Camomile, German chamomile, Hungarian chamomile, Wild chamomile, Blue chamomile, Scented mayweed: 7: §70; 10: §100 ‣ XII, 22

Chelidonium (Chelidonium majus) {curr.lat. *Chelidonium majus* L.} ■ Greater celandine: 10: §99 ‣ VI, 73

Cicoraceum (genus)[Cicoracea] {curr.lat. *Cichorium (genus)*} ■ Chicoraceous: 10: §98; 12: §132; 14: §151 ‣ [XIII, 1]

Cicorium {curr.lat. *Cichorium endivia* L.?;} ■ Chicory, Escarole: 7: §74; 12: §127 ‣ XIII, 2

Ciperus (Cyperus hortensis) {curr.lat. *Cyperus esculentus* L.} ■ Chufa Sedge, Tiger nut, Atadwe, Yellow nutsedge, Earth almond: 4: §40; 5: §50 ‣ IV, 63

Clematis {anc.gr. Κληματίς; curr.lat. *Clematis flammula* L.} ■ Clematis: 11, §114 ‣ XIV, 4

Colchicum {anc.gr. Κολχικόν ἐφήμερον; curr.lat. *Colchicum autumnale* L.} ■ Autumn crocus, Meadow saffron, Naked ladies: 10: §111 ‣ X, 49

Coronarium ("genus") ■ Crowned: 12: §128

Crocus {anc.gr. Κρόκος; curr.lat. *Crocus vernus* (L.) Hill} ■ Spring crocus, Giant crocus: 7: §70 ‣ X, 20

Cucumis {curr.lat. *Cucumis sativus* L.?} ■ Cucumber: 9: §90 ‣ V, 5

Cucumis sylvestris {anc.gr. Σίκυς ἄγριος; curr.lat. *Ecballium elaterium* (L.) A.Rich.} ■ Squirting cucumber, exploding cucumber: 10: §102 ‣ V, 6

Cucurbita (genus) {curr.lat. *Cucurbitacea (family)*} ■ Marrows: 7: §75; 9: §91; 10: §103; 11: §114 ‣ V, 2, 3, 7

Cyclaminus {anc.gr. Κυκλάμινος; curr.lat. *Cyclamen repandum* Sm.} ■ Spring sowbread: 1: §6; 10: §110, §111, 12: §131 ‣ IX, 38

Dentaria minor (Aconitum lycoctonum) {anc.gr. Ἀκόνιτον λυκοκτόνον; curr.lat. *Cardamine bulbifera* (L.) Crantz} ■ Coral root: 5: §51 ‣ XV, 33

Elleborum candidum {anc.gr. Ἐλλέβορος λευκός; curr.lat. *Veratrum album* L.} ■ False helleborine, White hellebore, European white hellebore, White veratrum: 12: §132, §133 ‣ XV, 27

Elleborum nigrum {anc.gr. Ἐλλέβορος μέλας; curr.lat. *Helleborus niger* L.} ■ Christmas rose, black hellebore: 12: §133 ‣ XV, 28

Elleborus (Helleborus (genus) {anc.gr. Ἐλλέβορος; curr.lat. *Helleborus viridis* L.; *Helleborus (genus)*} ■ Green hellebore; Hellebores: 7: §75; 12: §131 ‣ XV, 28

Eruca {curr.lat. *Eruca sativa* Mill.?} ■ Rocket, Arugula: 12: §132 ‣ VIII, 62

Ferula {anc.gr. Th. Νάρθηξ; curr.lat. *Ferula communis* L.?} ■ Giant Fennel: 3: §32; 4: §39; 12: §121, §128 ‣ VII, 2

Ferulaceum (genus) ■ Ferulaceous: 1: §6; 4: §39; 10: §95, §109; 12: §120, §128, §132; 14: §150 ‣ VII, 1

Ficus {curr.lat. *Ficus carica* L.?} ■ Common fig: 2: §17; 7: §75; 10: §110, §111 ‣ III, 1, 2

Ficus aegyptia {curr.lat. *Ficus sycomorus* L.?} ■ Sycamore fig, Fig-mulberry: 8: §82 ‣ III, 2, 28

Filix {anc.gr. Πτέρις; curr.lat. *Osmunda regalis* L.} ■ Royal Fern: 14: §148 ▸ XVI, 4
Foeniculum {curr.lat. *Foeniculum vulgare* Mill.?} ■ Fennel: 4: §40; 10: §95 ▸ VII, 9
Fragaria (genus) {curr.lat. *Fragaria (genus)*} ■ Strawberry plants; Strawberries: 10, §93, §100 ▸ XIV, 18
***Frumentum (genus)* [*Frumenta*]** ■ Cereals: 4: §42; 7: §74; 8: §79, §80, §82; 10: §93, §106, §109; 12: §127; 13: §140, §145 ▸ [IV, 41]
Frutex (genus) {anc.gr. Th. Θάμνος; } ■ Shrubs: 12: §118–§121, §124, §128; 13: §137, §138, §144; 14: §157 ▸ II, III
***Fungus (genus)* [*Fungi*]** ■ Mushrooms: 14: §145 ▸ XVI, 37, 40
Gladiolus {anc.gr. Ξίφιον; curr.lat. *Gladiolus italicus* Mill.} ■ Field gladiolus, Italian gladiolus Common sword-lily: 4: §41; 6: §60 ▸ V, 35
***Glandifer (genus); Glans* [*unguentaria*] [*Glandiferatum differentiae*]** ■ Glandiferous; Acorns: 8: §79; 10: §93, §97 ▸ [II, 2, 45]
Gramen {anc.gr. Ἄγρωστις; curr.lat. *Cynodon dactylon* (L.) Pers.} ■ Bermuda Grass Dhoob, dūrvā grass, Ethana grass, Dubo, Dog's tooth grass, Bahama grass, Devil's grass, Couch grass, Indian doab, Arugampul, Grama, Wiregrass, Scutch grass: 4: §40; 5: §50 ▸ IV, 57
Harundo (Arundo)(genus) {curr.lat. Arundo (genus)?} ■ Reeds: 3: §32; 4: §44 ▸ IV, 59
Hedera {curr.lat. *Hedera helix* L.?} ■ (Common) Ivy: 5: §51; 11: §114 ▸ III, 5
Hedera terrestris (Ederra terrestris) {curr.lat. *Glechoma hederacea* L.} ■ Ground-ivy; Gill-over-the-ground, Creeping charlie, Alehoof, Tunhoof, Catsfoot, Field balm, Run-away-robin: 5: §50 ▸ XI, 36
Helenium {anc.gr. Ἑλένιον; curr.lat. *Inula helenium* L.} ■ Elecampane, Horse-heal, Elfdock: 5: §48 ▸ XII, 34
***Helxina* [*Helxine*]** {anc.gr. Ἑλξίνη; curr.lat. *Soleirolia soleirolii* (Req.) Dandy?} ■ Baby's tears, Angel's tears, Peace in the home, Bits and pieces, Bread and cheese, Corsican creeper, Corsican curse, Friendship plant, Mind-your-own-business, Mother of thousands, Paddy's wig, Pollyanna vine: 11: §114 ▸ IV, 38
Herba (genus) {anc.gr. Πόα; } ■ Herbs: 7: §73; 10: §111; 11: §114; 12: §118–124, §125, §128, §129; 13: §137–138; 14: §157; ▸ IV–XVI
Hordeum (Pl.: Ordeum murinum) {curr.lat. *Hordeum murinum* L.} ■ Wall barley, False barley: 10: §95 ▸ IV, 58
Humirubus (Chamaebatus) {anc.gr. Χαμαίβατος; curr.lat. *Rubus caesius* L.} ■ Dewberry: 5: §50 ▸ XIV, 18
Hypericum {anc.gr. Ὑπερικόν; curr.lat. *Hypericum perforatum* L.} ■ Perforate St John's-wort: 12: §132 ▸ IX, 41
Hypocistis {anc.gr. Ὑποκιστίς; curr.lat. *Cytinus hypocistis* (L.) L.} ■ Cytinus, Hypocist, Yellow Cytinus: 14: §146 ▸ XV, 15
Ilex {curr.lat. *Quercus coccifera* L.?} ■ Kermes oak: 11: §116 ▸ II, 2
Intubus {curr.lat. *Cichorium intybus* L.?} ■ Belgian endive, Witloof: 8: §84 ▸ XIII, 2
Irio (Erysimum) {anc.gr. Ἐρύσιμον; curr.lat. *Sisymbrium officinale* (L.) Scop.} ■ Hedge mustard: 12, §127 ▸ VIII, 64
Iris {anc.gr. Ἶρις; Ξῖρις; curr.lat. *Iris graminea* L.; *Iris foetidissima* L.} ■ Grass-leaved flag, Grass leaved iris, Plum iris, Plum tart iris; Stinking iris, Gladdon, Gladwin iris, Roast-beef plant, Stinking gladwin: 4: §40 ▸ X, 37; 38
Iuncus acutus {curr.lat. *Juncus acutus* L.?} ■ Spiny rush, Sharp rush, Sharp-pointed rush: 11: §115; (12: §132) ▸ IV, 60, 70
Lactuca (genus) {curr.lat. *Lactuca (genus)*} ■ Lettuces: 12: §132 ▸ XIII, 19
Lactuca domestica {curr.lat. *Lactuca sativa* L.?} ■ Garden lettuce: 12: §133
Lactuca sylvestris {curr.lat. *Lactuca virosa* L.?} ■ Wild lettuce, Bitter lettuce, Opium lettuce, Poisonous lettuce, Tall lettuce, Great lettuce: 12: §133 ▸ VI, 57; VIII, 60; XIII, 13, 20
Laurus sylvestris (Pline: Tino) {curr.lat. *Viburnum tinus* L.} ■ Laurustinus, Laurustine, Laurestine: 3: §31; 9: §89, §90 ▸ II, 40
***Legumen (family)* [*Legumina*]** {curr.lat. *Fabaceae, Leguminosae (family)*} ■ Legumes: 6: §59, §62; 8: §77, §79, §80, §82; 10: §93, §97, §99; 11: §114; 12: §122, §127; 13: §141 ▸ [VI, 1]
Lenticula palustris {curr.lat. *Lemna trisulca* L.} ■ Star duckweed, Ivy-leaved duckweed: 14: §145 ▸ XVI, 35

Lichen (genus) {anc.gr. Λειχήν} ■ Lichens: 14: §145 ▸ XVI, 20
Ligustrum (genus) {curr.lat. *Ligustrum*} ■ Privets: 10: §108 ▸ III, 8
Lilium {curr.lat. *Convallaria majalis* L.?; *Lilium bulbiferum* L.?; *Lilium martagon* L.?} ■ Lily of the valley; Orange lily (Fire lily, Jimmy's Bane, Tiger lily); Martagon lily (Turk's cap lily): 7: §70 ▸ V, 34; X, 33
***Limus (Citrus) (genus)* [*Limon*]** {curr.lat. *Citrus (genus)*} ■ Citrus fruits: 9: §90 ▸ [III, 59]
Linum (genus) ■ Flaxes: 4: §46 ▸ XIV, 35
Lithospermum {anc.gr. Λιθόσπερμον; curr.lat. *Lithospermum officinale* L.} ■ Common gromwell, European stoneseed: 4: §46; 10: §94 ▸ XI, 8
Lychnis {anc.gr. Λυχνίς ἀγριά; curr.lat. *Agrostemma githago* L.; *Atocion armeria* (L.) Raf.} ■ Common corncockle; Sweet William catchfly: 10; §97–99, §104 ▸ VI, 43, 44
Malus (genus); (pomum) {curr.lat. *Malus (genus)*} ■ Apples; Apple trees: 7: §67; 8: §77, §81; 9: §90 ▸ III, 61
Malva (genus) {anc.gr. Μαλάχη; curr.lat. *Malva (genus); Malva moschata* L.?; *Malva sylvestris* L.?;} ■ Mallows; Musk mallow; Common mallow, Cheeses, High mallow, Tall mallow: 12: §121 ▸ XIV, 30, 34
Mandragora {anc.gr. Μανδραγόρα; curr.lat. *Mandragora officinarum* L.; *Mandragora autumnalis* Bertol.} ■ Mandrake; Autumn mandrake: 9: §91; 10: §93, §110 ▸ V, 24
Marrubium (Prasium) {anc.gr. Πράσιον; curr.lat. *Marrubium vulgare* L.} ■ White horehound, Common horehound: 4: §44; 10: §96; 14: §151 ▸ XI, 26
Melica {curr.lat. *Melica altissima* L.?} ■ Melic, Melic grass: 5: §51 ▸ IV, 53
Melopepon {curr.lat. *Cucumis melo* L.?} ■ Melon: 8: §77; 9: §90 ▸ V, 5
Mercurialis (Linozostis) {anc.gr. Λινόζωστις; curr.lat. *Mercurialis annua* L.; *Mercurialis perennis* L.} ■ Mercury (French mercury); Dog's mercury: 7: §73 ▸ VIII, 2
Milium {curr.lat. *Panicum miliaceum* L.?} ■ Proso millet, Broomcorn millet, Common millet, Hog millet, Kashfi millet, Red millet, White millet: 10: §109 ▸ IV, 51
Milium indicum {curr.lat. *Sorghum bicolor* (L.) Moench?} ■ Sorghum, Great millet, Durra, Jowari, Jowar; Milo: 7: §72; 10: §96; §109; ▸ IV, 53
Morus {curr.lat. *Morus nigra* L.?} ■ Black mullberry, Blackberry: 10: §100 ▸ III, 3
Muscus (Iuba) (genus) {curr.lat. *Usnea (genus)?*} ■ Old man's beards, Beard lichens: 11: §116 ▸ XVI, 21
***Napum* [*Napus*]** {curr.lat. *Brassica rapa* L. subsp. *rapa*?} ■ Turnip: 12: §131 ▸ [VII, 42; VIII, 57]
Narcissus (genus) {anc.gr. Νάρκισσος; curr.lat. *Narcissus (genus)*} ■ Daffodils, Narcissus, Jonquils: 10: §111 ▸ X, 22, 23, 24
Nasturtium {anc.gr. Κάρδαμον; curr.lat. *Lepidium sativum* L.} ■ Cress, Garden Cress: 14, §151 ▸ VIII, 70
Nepa {anc.gr. Th. Σκορπίος; curr.lat. *Ulex europaeus* L.?} ■ Gorse, Common gorse, Furze, Whin: 11: §115 ▸ III, 37
Nux; Nucleus {curr.lat. *Juglans (genus)*?} ■ Walnut; Walnut tree: 6: §60; 8: §77, §79; 9: §91 ▸ II, 5
Nux avellana (Nux pontica) {anc.gr. Καρύα ποντική; curr.lat. *Coryllus avellana* L.} ■ Common Hazel; Hazelnut: 7: §72; 10: §97 ▸ II, 6
***Nux unguentaria (Pl.)* [*Carya*]** {anc.gr. Καρύα; curr.lat. *Moringa oleifera* Lam.?; *Myristica fragrans* Houtt.?} ■ Moringa, Drumstick tree, Horseradish tree, Ben oil tree, Benzolive tree; Nutmeg: 10, §97 ▸ [II, 5]
Oenanthe {anc.gr. Οἰνάνθη; curr.lat. *Filipendula ulmaria* (L.) Maxim.; *Filipendula vulgaris* Moench} ■ Meadowsweet, mead wort, Queen of the meadow, Pride of the meadow, Meadow-wort, Meadow queen, Lady of the meadow, Dollof, Meadsweet, Bridewort; Dropwort, Fern-leaf dropwort: 12: §132 ▸ XIV, 14; XIV, 13
Olea {anc.gr. ἐλαία; curr.lat. *Olea europea* L.?} ■ Olive tree; olive: 1: §11; 3: §35; 5: §49; 7: §67; 9: §90 ▸ II, 35
Olus ■ Vegetable varieties, Vegetable plants, Vegetables: 12: §122, §127
Ononis {anc.gr. Ὀνωνὶς; Ἀνωνὶς; curr.lat. *Ononis spinosa* L.} ■ Spiny restharrow: 11: §115 ▸ VI, 21
Opontia (genus) {curr.lat. *Opuntia (genus)*} ■ Prickly pears, Nopals, Paddle cactus: 10: §110
Orminum (Horminum sylvestre) {anc.gr. Ὅρμινον; curr.lat. *Melittis melissophyllum* L.} ■ Bastard balm: 7: §75 ▸ XI, 29

Ornithogalum {anc.gr. Ὀρνιθόγαλον; curr.lat. *Ornithogalum umbellatum* L.; *Loncomelos narbonensis* (L.) Raf.; *1.1.1 Loncomelos pyrenaicus* (L.) Hrouda} ■ Garden star-of-Bethlehem, Grass lily, Nap-at-noon, Eleven-o'clock lady; Narbonne star-of-Bethlehem, Pyramidal star-of-Bethlehem, Southern star-of-Bethlehem; Prussian asparagus, Wild asparagus, Bath asparagus, Pyrenees star of Bethlehem, Spiked star of Bethlehem: 7: §75 ▸ X,10, 11

Orobancha (Orobanche) {anc.gr. Ὀροβάγχη; curr.lat. *Orobanche minor* Sm.} ■ Hellroot, Common broomrape, Lesser broomrape, Small broomrape, Clover broomrape: 14: §146 ▸ VIII, 39

Oxycedrus {anc.gr. Th. Ὀξύκεδρος; Ginepro (dios.); curr.lat. *Juniperus oxycedrus* L.} ■ Cade, Cade juniper, Prickly juniper, Prickly cedar, Sharp cedar: 7: §73 ▸ III, 58

Oxys (Pl.: Alleluya) {curr.lat. *Oxalis corniculate* L.} ■ Creeping woodsorrel, Procumbent yellow sorrel, Sleeping beauty: 10: §101 ▸ XV, 2

Palma (Phoenix); palmeus {anc.gr. Φοῖνιξ; curr.lat. *Phoenix dactylifera* L.} ■ Date, Date palm; palm nut: 4: §42; 6: §60; 7: §73; 8: §79; 10: §110 ▸ II, 46

Panicum (genus) {curr.lat. *Panicum*} ■ Panicgrasses: 10: §109 ▸ IV, 52

Papaver {curr.lat. *Papaver rhoeas* L.?; *Glaucium corniculatum* (L.) Rudolph?} ■ Common poppy (Corn poppy, Corn rose, Field poppy, Flanders poppy, Red poppy); Blackspot hornpoppy (red horned-poppy): 7: §70, §74; 10: §97 ▸ VI, 74; XV, 10

Papyrus {anc.gr. Πάπυρος; curr.lat. *Cyperus papyrus* L.} ■ Papyrus, Papyrus sedge, Paper reed, Indian matting plant, Nile grass: 12: §132 ▸ IV, 65

Pentafillus (Pentaphyllon; Quinquefolium) {anc.gr. Πεντάφυλλον; curr.lat. *Potentilla reptans* L.} ■ Creeping cinquefoil, European cinquefoil, Creeping tormentil: 5: §50 ▸ XIV, 49

Peonia (genus) {anc.gr. Παιωνία; curr.lat. *Paeonia (genus)*} ■ Peonies, Paeonies: 10: §104

Periploca {anc.gr. Περιπλοκή; curr.lat. *Periploca graeca* L.} ■ Silkvine: 11: §114 ▸ III, 42

Persicum [Persica] {anc.gr. Περσική; curr.lat. *Prunus persica* (L.) Batsch} ■ Peach tree; Peach: 8: §79, §82; 9: §90; 10: §103 ▸ II, 17

Phaselus (genus) {anc.gr. Φάσηλος; curr.lat. *Phaseolus (genus)*} ■ Beans; Broad beens, Fava beans: 10: §101 ▸ VI, 14

Phyllitis (Lingua cervina) {anc.gr. Φυλλῖτις; curr.lat. *Asplenium scolopendrium* L.} ■ Hart's-tongue, hart's-tongue fern: 14: §148 ▸ XVI, 7.

Pinus (genus); Pineus ■ Pines; Cones, Fir cones; Pine nuts, pine kernels: 1: §10; 3; §31; 4: §46; 8: §77, §79; 10: §93 ▸ III, 52

Pirus {curr.lat. *Pyrus communis* L.?} ■ Pear tree; Pear: 7: §67; 8: §77, §81; 9: §87 ▸ III, 62

Pisum {curr.lat. *Pisum sativum* L.?} ■ Pea: 10: §101 ▸ VI, 4

Polemonia {anc.gr. Πολεμώνιον; curr.lat. *Silene vulgaris* (Moench) Garcke} ■ Bladder campion, Maidenstears: 7: §74; 10: §98, §104 ▸ X, 50

Polygonum {anc.gr. Πολύγονον; curr.lat. *Plygonum aviculare* L.; *Polygonum maritimum* L.} ■ Common knotgrass, Prostrate knotweed, Birdweed, Pigweed, Lowgrass; Sea knotgrass: 12: §132 ▸ IV, 36

Populus (genus) {curr.lat. *Populus (genus)*; *Populus alba* L.; *Populus nigra* L.} ■ Poplars; White poplar; Black poplar: 10: §98 ▸ III, 45

Porrum {curr.lat. *Allium ampeloprasum* L.?} ■ Wild leek, Broadleaf wild leek: 5: §52; 8: §84 ▸ X, 12; XVI, 25

Portulaca {curr.lat. *Portulaca oleracea* L.?} ■ Common purslane, Duckweed, Little hogweed, Pursley: 12: §127 ▸ VI, 61

Prunus {curr.lat. *Prunus domestica* L.?} ■ European plum; Plum: 7: §67; 10: §103 ▸ II, 15

Punica {curr.lat. *Punica granatum* L.?} ■ Pomegranate tree; Pomegranate: 1: §11; 5: §49; 7: §75; 9: §89, §90, §91; 10: §98, §103 ▸ III, 60

Ranunculus ("quiddam ranunculus") {curr.lat. *Ranunculus repens* L.} ■ Creeping buttercup, Creeping crowfoot, Sitfast: 5: §50 ▸ XIV, 3

Ranunculus (genus) {curr.lat. *Ranunculus (genus)*} ■ Buttercups: 7: §70; 10: §100; 12: §131, §132 ▸ XIV, 2, 3, 4, 5, 6, 7

Raphanus Montanus (Raphano) {anc.gr. Θλάσπις; curr.lat. *Armoracia rusticana*; G.Gaertn., B.Mey. & Scherb.} ■ Horseradish: 4: §40; 5: §48 ▸ VIII, 78
Rapum **[_Rapa_]; [_Rapus_]** {curr.lat. *Brassica napus* var. *napobrassica* (L.) Rchb.?;} ■ Swede, Rutabaga: 12: §131 ▸ [VIII, 58]; [VIII, 57]
Rhododap[h]ne {anc.gr. Ῥοδοδάφνη; curr.lat. *Nerium oleander* L.} ■ Oleander, Nerium: 10: §98 ▸ III, 41
Rhus {curr.lat. *Cotinus coggygria* Scop.?} ■ European smoketree, Eurasian smoketree, Smoke tree, Smoke bush, Venetian sumach, Dyer's sumach: 10: §110 ▸ II, 39
Ricinus (Cici) {anc.gr. Κίκι; Th. Κροτών; curr.lat. *Ricinus communis* L.} ■ Castor bean, Castor oil plan: 7: §73; 10: §101 ▸ IX, 19
Robur {curr.lat. *Quercus robur* L.?} ■ Common oak, Pedunculate oak, European oak, English oak: 11: §116 ▸ II, 2
Rosa (genus) {anc.gr. *Th. Ῥόδον*; curr.lat. *Rosa (genus)*} ■ Roses; Rose bushes: 7: §67, §74; 10: §100 ▸ III, 10
Rosmarinus (candidus rosmarinus) **[_Rosmarinum_]** {curr.lat. *Rosmarinus officinalis* L.?} ■ Rosemary: 10: §95 ▸ VII, 44
Rubus {anc.gr. Βάτος ἰδαία; curr.lat. *Rubus idaeus* L.} ■ European red raspberry; Raspberry: 10: §100 ▸ III, 11
Rubus caninus (Cyna, Cynus) {anc.gr. χυνάζ; curr.lat. *Rosa canina* L.?} ■ Dog rose: 11: §116 ▸ III, 10
Ruscus {anc.gr. Ὀξυμυρσίνη; curr.lat. *Ruscus aculeatus* L.} ■ Butcher's-broom: 6: §60; 8: §83 ▸ V, 28
Ruta (genus) {curr.lat. *Ruta (genus)*} ■ Rues: 12: §120 ▸ XV, 14
Salix (genus) {curr.lat. *Salix (genus)*; *Salix alba* L.} ■ Willows; White willow: 3: §35 ▸ III, 46
Sampsucus (Sampsuchum, Persa) {anc.gr. Σάμψουχον; Σάμψυχον; curr.lat. *Origanum vulgare* L.} ■ Wild marjoram, Oregano: 12: §120 ▸ XI, 53
Satyrium (Satyrium erithronium) {anc.gr. Σατύριον ἐρυθρόνιον; curr.lat. *Iris tuberosa* L.} ■ Snake's-head, Snake's-head iris, Widow iris, Black iris, Velvet flower-de-luce: 4: §41 ▸ X, 45
Sesamoidis (Sesamoides) {anc.gr. Σησαμοειδές; curr.lat. *Reseda lutea* L.} ■ Yellow mignonette, Wild mignonette: 12: §132 ▸ IX, 34
Sesamum **[_Sesama_]** {anc.gr. Σήσαμον; curr.lat. *Sesamum indicum* L.?} ■ Sesame: 12: §127 ▸ [VIII, 79; IX, 33]
Siler (Pl. Silio) {anc.gr. Εὐώνυμος; curr.lat. *Euonymus europaeus* L.} ■ Spindle, European spindle, Common spindle: 14: §151 ▸ III, 50
Siliqua {anc.gr. Κερωνία; curr.lat. *Ceratonia siliqua* L.} ■ Carob tree; Carob: 8: §82; 10: §101 ▸ III, 27
Sinapi {anc.gr. Σίναπι; Σίνηπι; curr.lat. *Brassica nigra* (L.) W.D.J.Koch} ■ Black mustard: 6: §60 ▸ VIII, 61
Smilax (Smilax laevis; Convolvulus maior) {anc.gr. Σμῖλαξ λεῖα; curr.lat. *1.1.2 Convolvulus sepium* L.} ■ Hedge bindweed, Rutland beauty, Bugle vine, Heavenly trumpets, Bellbind, Granny-pop-out-of-bed: 7. §70 ▸ IX, 25
Spartum **[_Sparton_]** {curr.lat. *Lygeum spartum* L.?} ■ Esparto grass, Cord grass, Albardine: 10: §101 ▸ [III, 35]
Solanum (Solanum arborescens) {curr.lat. *Solanum pseudocapsicum* L.} ■ Nightshade, Jerusalme cherry: 10: §93 ▸ V, 23
Suber {curr.lat. *Quercus suber* L.?} ■ Cork oak: 3: §37 ▸ II, 2
Suffrutex (genus) {anc.gr. Φρύγανον; } ■ Undershrubs: 5: §49; 12: §118, §119, §120, §121, §124, §128; 13: §137, §138; 14: §157 ▸ IV–XVI
Taxus {anc.gr. Σμῖλαξ; curr.lat. *Taxus baccata* L.} ■ Yew: 7: §73 ▸ III, 34
Testiculus (genus) **[_Testiculorum genera_]** {anc.gr. Ὄρχις; curr.lat. *Orchis (genus)*} ■ Orchids: 14: §146 ▸ X, 46
Thaliethrum (thalietron) {anc.gr. Θαλίητρον; curr.lat. *Thalictrum aquilegiifolium* L.} ■ Siberian columbine, Meadow-rue, Columbine meadow-rue, French meadow-rue, Greater meadow-rue: 7: §71 ▸ XI, 2
Tithymalus (genus) {anc.gr. Τιθύμαλος; curr.lat. *Euphorbia (genus)*} ■ Euphorbias, Spurges: 2: §17; 4: §46; 10: §110; 12: §121; 14: §151 ▸ IX, 5, 11, 15
Tribulus {anc.gr. Th.: Τρίβολος; curr.lat. *Tribulus terrestris* L.} ■ Goat's-head, Bindii, Bullhead, Burra gokharu, Bhakhdi, Caltrop, Small caltrops, Cat's-head, Devil's eyelashes, Devil's-thorn, Devil's-weed, Puncture vine, Tackweed: 8: §83; 10: §94; 11: §115 ▸ IV, 27

Triticum {curr.lat. *Triticum aestivum* L.?} ■ Common wheat, Bread wheat: 3: §32; 4: §42; §43; 6: §58, §62; 8: §80; 10: §96; 12: §131 ▸ IV, 42

Ulmus (genus) {curr.lat. *Ulmus (genus)*} ■ Elms: 11: §116 ▸ II, 9

Urtica (Acalyphe) {anc.gr. Ἀκαλύφη; curr.lat. *Urtica dioica* L.?; *Urtica Pilulifera* L.?} ■ Common nettle, Stinging nettle, Nettle leaf; Roman nettle: 7: §73 ▸ IV, 15

Verbenaca {anc.gr. Περιστερεών; curr.lat. *Verbena officinalis* L.; *Lycopus europaeus* L.;} ■ Common vervain, Common verbena; Gypsywort, Bugleweed, European bugleweed, Water horehound: 10: §110 ▸ XI, 32, 33

Vesicaria (Pl.: Halicacabum), Faba inversa; Solanum vesicarium {anc.gr. Στρύχνος; Ἀλικακάβος; curr.lat. *Cardiospermum halicacabum* L.; *Physalis alkekengi* L.} ■ Baloon plant, Love in a puff; Bladder cherry, Chinese lantern, Japanese-lantern, Strawberry groundcherry, Winter cherry: 7: §74 ▸ V, 20; IX, 24

Viola (genus) {curr.lat. *Viola (genus)*} ■ Violets: 10: §110 ▸ VI, 45; VIII, 80; IX, 39, 40

Vitis (vua) {curr.lat. *Vitis vinifera* L.?} ■ Grapevine; grapes: 1: §11; 2: §17; 4: §42; 5: §49, §50; 8: §82; 9: §87, §90; 10: §105, §108, §110; 11: §114; 12: §132 ▸ III, 18

Zea {curr.lat. *Triticum aestivum* subsp. *spelta* (L.) Thell.?} ■ Spelt, Dinkel wheat, Hulled wheat: 10: §95 ▸ IV, 43, 44

Thematic Index

This index includes the main terms and concepts from book 1. It is divided into an *index nominum* and an *index rerum*.

The *index rerum* is organized thematically; entries are grouped by category. Grammatically related terms are indexed together (e.g., a noun and a derived adjective). Latin terms or phrases are italicized and followed by the English translation in brackets. References are given as chapter number followed by paragraph(s).

Index nominum

Dioscorides: 12: §134.
Theophrastus: 12: §134.

Index rerum

Analogies

With animals: 1: §1, §2, §3, §6, §7, §8, §13, §14; 2: §15, §16, §17, §18, §19, §24; 3: §27, §28, §29; 4: §38; 5: §48, §53; 6: §56; §57, §58, §59, §60, §61; 7: §64, §65; 8: §78, §80; 9: §90; 11: §114; 13: §143; 14: §145, §146.
With artificial objects: 2: §20, §22, §2; 10: §109.

Animals

animal (animal): 1: §1–3, §6–8, §13–14; 2: §15–19, §24; 3: §27–29; 4: §38; 5: §48, §53; 6: §56–60; 7: §65; 8: §78; 11: §114; 13: §143; 14: §145–146.
avis (birds, including butterfly): 2: §21; 7: §78.
insectum (insects): 5: §48; 7: §78.
ovum (egg): 3: §28; 5: §53; 6: §57–58, §61; 8: §78, §80.
piscis (fish): 8: §78.

Aristotelianism

accidens (accident): 13: §135, §143; 14: §154, §155.
actus (actual-ity): 1: §8; 4: §40; 5: §48.
anima (soul): 1: §1, §2, §5, §8, §9, §10, §11; 2: §21; 3§27; 13: §136, §137, §143; 14: §145.
facultas (faculty): 1: §2, §9, §10; 3: §27; 13: §135.
forma (form): 1: §2; 3: §30; 6: §57; 12: §127, §129; 13: §135, §136, §143.
natura (nature):
- ... of the plant: 1: §1, §3, §4, §13; 2: §21, §22, §24; 3: §35, §36, §37; 4: §40, §41; 5: §48; 6: §55, §60, §63; 7: §69; 9: §88, §91; 10: §93; 12: §119, §120, §131; 13: §137, §142; 14: §146, §147, §156.
- ... as a principle: 2: §26; 4: §40, §41; 5: §50; 6: §55; 7: §69, §74; 8: §78, §79, 9: §86, 10: §96, §97; 11: §114; 13: §143; 14: §149.
- *secundum naturam* (in accordance with nature; conforming to nature): 3: §30, §32, 12: §133.

necesse est, ex necessitate (idea of necessity): 1: §5, §7; 2: §18, §23, §25; 3: §28; 6: §56, §57, §61; 7: §65, §67; 8: §82; 11: §114, §116; 12: §126, §130; 13: §136.
per se (in itself): 5: §53; 14: §154, §155, §156.
potentia (potential-ity): 4: §40; 5: §48.
materia (matter): 3: §28; 6: §57; 7: §67, §69, §73, §74; 8: §80, §82, §84; 9: §88, §89; 13: §136.
- *humidi, humilior materia*: 1: §5; 3: §32; 4: §41; 13: §140.

Quaestiones peripateticae (*Peripatetic questions*): 1: §8; 6: §57; 13: §135.
substantia (substance): 1: §2, §4, §14; 2: §22; 3: §27, §32, §34, §35, 6: §56, §61, §63; 7: §65, §67, §72; 8: §82, §84; 9: §91; 10: §95, §98, §101, §110; 12: §118, §120; 13: §135, §137, §143; 14: 56.

Classification

arbor (tree): 1: §5, §12; 2: §15; 3: §30–32, §35–37; 4: §38, §42, §46; 5: §49, §51; 8: §82; 9: §88, §91; 10: §93, §100, §110, §111, §112; 11: §114, §116; 12: §118, §119, §124, §128, §133; 13: §137, §138; §140, §144; 14: §153, §157.
cibus, mandere (plants as food): 8: §77; 9: §87; 10: §101.
distinguere, *diuidere*, *diuisio* (division; as a method): 6: §57; 8: §79; 13: §138, §139, §142; 14: §149, §157.
domesticus (cultivated): 9: §90, §91; 12: §133.
figura (shape): 6: §55; 8: §83; 12: §132; 14: §149, §150, §153.
frumentum (cereals): 4: §42; 7: §74; 8: §80, §82; 10: §93, §109; 12: §127; 13: §141; 14: §146.
frutex (shrub): 12: §118–121; §124, §128; 13: §137, §138; §144; 14: §157.
genus (genus): 1: §1; 2: §17, §19; 3: §37; 4: §46; 5: §51; 7: §66, §72; 9: §90; 10: §98, §111; 11: §114; 12: §128, §129, §131, §132, §133; 13: §143; 14: §145, §149, §151.
herba (herb): 7: §73; 10: §111; 11: §114; 12: §118, §119, §121, §122, §123, §124, §125; §128, §129; 13: §137, §138; 14: §157.
legumen (legumes): 6: §59, §62; 8: §77, §79, §80, §82; 10: §93, §97, §99; 11: §114; 12: §122, §127; 13: §141.
numerus (number): 1: §2; 4: §46; 6: §55; 13: §138; 14: §149, §150.
olus (vegetable): 12: §122, §127.
situs (position): 14: §149, §152.
species (species): 1: §5, §9, §10; 12: §126, §130, §132, §133; 13: §135, §136, §143; 14: §147, §149, §154, §156, §157.
suffrutex (undershrub): 5: §49; 12: §118, §119, §120, §121; §124, §128; 13: §137, §138; 14: §157.
sylvester (wild plants): 9: §90, §91; 12: §133.
usus (usage): 12: §134; 13: §135.
vis, facultas, proprietas (medicinal properties): 12: §123, §134; 13: §135; 14: §156.

Finalism

datur, *data est* (endowed with): 1: §5; 2: §16, §17, §19; 3: §29, §30, §31, §33, §34; 4: §42, §43; 6: §61, §63; 7: §64, §65, §67; 8: §79, §80, §83; 9: §90; 10: §93, §96, §103, §107, §113, §116; 13: §136, §137, §139, §140, §144; 14: §154.
gratia (in order to, on account of, etc.): 1: §2; 2: §16, §19; 3: §27, §29, §30, §31, §33, §34; 4: §42; 6: §63; 8: §78; 9: §86; 10: §107; 11: §114, §116; 13: §136, §137, §140, §144; 14: §154.
Idea of perfection, maturity, completion: 1: §5, §6; 2: §15; 3: §29, §31; 6: §55, §61; 7: §64; 8: §81, §84; 9: §86, §87, §91; 10: §94, §101, §112; 12: §122; 13: §140, §143; 14: §145, §146, §147, §149.
rationi consonum (in accordance with reason): 1: §13; 9: §88.
recta ratione (for a good reason): 2: §16; 3: §31, §34; 10: §100.

Medicine

cathaplasma (poultice): 9: §86.
herbarius (herbalist): 12: §123.
medicus (physician): 9: §86; 14: §156.

Physics

aereus (air): 6: §60.
aqueus, *aqua* (water-y): 6: §60; 7: §69.
congelatio (freezing): 2: §25.
decidere (idea of condensation): 7: §66.
euanescere (evaporation): 8: §78.
frigidus, *frigor* (cold): 2: §17, §25.
grauis, *grauitas* (heavy, heaviness): 2: §23; 12: §119.
humidum, *humectus* (moist-ure): 2: §21; 6: §60–61; 9: §86.
humidus (wet): 6: §56, §60; 7: §74; 9: §90.
humilis (of the) soil): 1: §6; 3: §32; 4: §41, §42; 13: §140, §144; 14: §153.
ignis (fire): 6: §61.
leuis, *leuitas* (light, lightness): 1: §7; 2: §23; 12: §132; 13: §139 (3: §32= *tenuis*).
magnes (magnet): 2: §20.
siccus (dry, dried): 2: §21, §25; 3: §32; 6: §60–61; 8: §84; 9: §90–91; 10: §98; 6: §62; 7: §69; 9: §91.
sorbere, *bibula natura* (idea of absorption): 2: §21, 22.
terra (earth): 1: §3, §6, §9, §10, §23; 5: §48, §50, §51, §53; 6: §58, §61; 7: §69.

urere (burn): 3: §30.
warm: see *calidus.*

Plant functions (physiology)

calor, calidus, caliditas (heat): 1: §7; 2: §15, §16, §17, §22, §23, §24, §25; 3: §29; 6: §56, §60; 7: §73; 8: §84; 9: §86, §87.
coelum (climate): 14: §155.
coquere, coctio (coction): 6: §61; 7: §66; 9: §86; 10: §101.
crescere, germinare (growth): 1: §1; 2: §15; 3: §27, 29, 37; 4: §38, §46; 6: §55, §57, §61; 10: §111; [12: §139].
dura et lignose ([becoming] hard and woody, i.e., lignification): 1: §14.
egere (desiring): 5: §51.
excrementum (excrement): 1: §3, §6; 6: §56; 8: §84; 9: §90; 11: §116.
foemina (female): 6: §57; 7: §73.
foetus (fetus): 1: §5; 3: §28, §29; 5: §53; 7: §64; 8§80; 9: §90.
fructificatio, fructificare (fructification): 4: §42; 6: §55; 7: §73, 8: §82; 10: §107, §110, §111, §112, §113; 13§140, §142, §143, §144; 14§154.
generare, nascere (reproduction, generation, birth): 3: §28, §37; 4: §38, §39, §40, §41, §44; 5: §53; 6§58; 13: §140; 14: §145.
- with the idea of spontaneous generation: 4: §38; 14: §145.
genitura (sperm): 6: §57, §59, §60.
germinatio, germinare (development, germination): 2: §24; 3: §27, §28, §29, §30, §31, §33, §34, §36, §37; 4: §38, §39, §40, §41, §42, §43, §45, §46, §47; 5: §50; 6: §55, §61; 7: §64; 8: §77, §79, §85; 10: §110, §112.
halitus, exhalation (steam; exhalation): 7: §65, §66, §69, §70, §73.
humor ([vital] liquid): 1: §14; 2: §17, §18, §19, §20, §21, §22, §23, §24, §25; 3: §30; 5: §53; 6: §61, §62; 7: §69; 8: §78, §84; 9: §86, §87, §90, §91; 10: §102.
inoculatio, infere (graft): 3: §30, §36, §37; 4: §42; 9: §88; 11: §115.
mas (male): 3: §28; 6: §57; 7: §73.
morbus (disease, degeneration): 2: §25; 7: §72; 14: §146.
mouere, motus (movement): 1: §1, §2; 2: §16, §19, §23; 13: §143.
nutrition, alimentum (food [for the plant]): 1: §3, §5, §7, §9; 2: §15, §16, §17, §18, §19, §20, §21, §26; 3: §28, §35, §37; 4: §48; 6: §56, §58, §61; 7: §74; 8: §81, §82, §84; 13: §137; 14: §154.
opus (operation): 1: §5, §9; 2: §16; 3: §27, §31; 13: §136, §137, §139, §140, §141, §142, §143.
per sobolem (by sobole, vegetative [reproduction]): 4: §41; 5: §50, §51, §52, §53; 6: §55.
perire, interire, emori, enecare, extinguere, mori (dying): 1: §10; 2: §25; 3: §36, §37; 4: §39, §41, 9: §91; 12: §122.
propagare, propagatio (propagation): 1: §5, §11; 3: §29; 4: §41; 5: §48, §53; 6: §55; 14; §148.
putritudo (rotting, putrefaction): 4: §38, 7: §69; 9: §90, §91; 14: §145.
radicula geniculata, moles (bulblet): 5: §51, §52.
senescare, senectus, senium, vetustus (aging): 1: §14; 2: §25; 3: §32; 4: §41; 5: §49; 11: §116.
sentire, sensus (sensation): 1: §1, §2; 2: §16, §19, §20; 3: §28; 11: §114; 13: §143.
sol (sun): 2: §29–30; 6: §61; 7: §66–67, §69; 9: §87, §89, §91; 10: §109.
spiritus (breath): 2: §15, §16, §18; 6: §56, §57, §59; 7: §65, §69.
sterilis (sterile): 7: §73; 14: §147.
velle (wanting): 9: §90.
venter (abdomen): 1: §3, §7; 2: §18.
viscera (viscera): 1: §2, §13.
viviradix (offset plant): 5: §50.
vmbra (shade): 3: §31.

Plant parts (morphology and anatomy)

aculeus (thorn): 11: §115, 12: §128.
alum (stipule, wing): 4: §42, §45; 5: §51; 10: §110; 11: §114.
amentum (catkin): 7: §72.
calyx (calyx): 7: §74, §75; 8: §84.
capreolus (tendrils): 11: §115; 13: §139.
cartilagineus (cartilaginous): 6: §60; 8: §79, §80; 10: §101, §103.
caulis (stem): 1: §5, §6, §7, §9, §11, §12, §13, §14; 2: §17, §18; 3: §32, §34, §35; 4: §438, §39, §41, §42, §43, §44, §46; 5: §48, §50, §51, §52; 7: §74, §75; 8: §82; 10: §94, §95, §100, §106, §108, §110, §111;

11: §114, §115; 12: §120, §124, §128, §132; 13: §139; 14: §147, §154.
cerebrum (brain): 1: §13, §14.
concameratio (corolla): 7: §66.
conceptaculum (receptacle): 6: §55; 7: §74; 14: §146, §151, §152, §153.
cor (heart): 1: §7, §8, §13; 2: §15, §18; 3: §29, §34; 4: §38, §39, §40, §42; 5: §48; 6: §56, §58, §61, §62.
corpulentia (body): 6: §57, §58, §61, 62; 12: §119.
– *corpusculum* (small body): 7: §70.
corpus (body): 1: §3, 4, §8, §12; 2: §15; 3: §32, §34, 35; 6: §59; 7: §72, §74; 8: §77, §78; 9: 6, §87, §88; 10: §93, §100; 11: §114; 12: §121.
cortex (bark): 1: §12, §13; 3: §30, §31, §32; §34, §35, §36, §37; 4: §42; 5: §53; 6: §61, §62; 7: §64, §68, §74, §75; 8: §78, §79, §80, §81, §82, §83, §84, §85; 9: §88, §89, §90; 10: §93, §94, §95, §98, §100.
divisio, *dividere* (division [in the physical sense]): 4: §38, §39, §40, §41; 5: §48, §50; 6: §55, §62; 7: §70, §74; 8: §79; 10: §97, §104, §105, §108, §110.
elementum (idea of germ): 6: §59.
floccus (flake, anther): 7: §70, §71, §72.
flos (flower): 6: §55; 7: §64, §65, §66, §67, §68, §69, §70, §71, §72, §73, §74, §75, §76; 8: §84; 9: §89; 10;§94, §95, §96, §97, §98, §99, §104, §105, §107, §108, §110, §111; 12: §128, §132, §133; 14: §148, §151, §152, §153.
flosculus (little flower): 7: §70; 10: §98, §106.
folium (leaf – including petals): 2: §17; 3: §30, §31, §32, §33, §34, §37; 4: §42, §44, §45, §46; 5: §51, §52; 6: §61, §62; 7: §67, §68, §69, §70, §71, §72, §74; 8: §84; 10: §95, §96, §106, §107, §110, §111; 11: §114, §115; 12: §123, §124, §127, §128, §132, §133; 13: §139; 14: §148, §154.
fructus (fruit): 1: §5; 2: §17, §23, §24; 3: §29, §30, §31, §34, §37; 4: §42, §47; 5: §50; 7: §65, §67, §70, §72, §73, §74, §75, §76; 8: §82; 9: §82, §87, §88, §90, §91; 10: §100, §101, §102, §104, §106, §107, §108, §109, §110, §113; 11: §114; 12: §128; 13: §140, §142; 14: §150, §152, §153, §154, §155.
gemma (bud): 3: §30; 4: §45.
geniculus (node): 4: §40, §43, §44, §46.
geniculus (sheath): 4: §40, §43, §44, §46; 5: §50.
germen (shoot): 1: §9, §10, §11, §12, §13, §14; 2: §23; 3: §27, §28, §29, §30, §31, §32, §34, §35, §37; 4: §38, §39, §40, §41, §42, §44, §45, §46; 5: §48, §49, §50, §53, §54; 6: §58, §61, §62; 7: §67; 8: §82; 9: §88, §89; 10: §93, §98, §107, §111; 11: §114; 12: §127, §131; 13: §137, §138.
idea of color: 6: §59; 7: §69, §74; 8: §84; 9: §89; 12: §128, §132; 14: §150, §155.
idea of scent, fragrance: 7: §65; 12: §128; 14: §155.
idea of taste: 2: §19; 9: §87, §89; 14: §156.
individuum, indivisum, indivisibilis, (undivided): 1: §10; 4: §38; 7: 0; 14: §147
intermedium qua radix germini coniungitur (idea of root collar): 1: §13, §14; 4: §39.
involucrum (husk): 7: §64, §74; 9: §86; 10: §94, §96, §97; 14: §151.
lanugo, *lanugineus* (down, downy): 7: §71, 10: §98; 14: §148.
liber (bast): 3: §32; 10: §100.
lignum (wood, woody): 1: §12, §14; 3: §34, §35, §36, §37; 4: §41; 8: §82; 9: §88; 12: §119, §124; 13: §137.
medulla (pith): 1: §12, §13, §14; 2: §18; 3: §34, §35, §36, §37; 4: §42; 6: §56, §59, §60; 7: §64; 8: §79, §80, §82, §83.
membrana, *membranaceus* (membrane, membranous): 7: §64; 8: §79, §80, 9: §90, §91; 10: §101.
– *membranula* (little membrane): 10: §106.
muscus (moss): 7: §71; 11: §116.
nervulus (nervures): 8: §82.
– *nerviculum* (fibrils): 10: §99.
– *nervus*, *neruosus* (fiber, fibrous): 1: §2; 2: §19; 4: §43; 9: §88; 10: §110; 12: §131; 13: §139.
os (bone, bony): 6: §60; 8: §79, §80.
os (stone): 1: §2; 6: §60; 8: §79.
pappus (pappi): 10: §98.
peciolus (radicle): 6: §61; 8: §81.
pediculus (peduncle): 3: §30, §32; 4: §42; 10: §102, §110; 11: §114.
pericarpium (pericarp): 8: §77; 9: §87, §88, §89, §90, §91, §91; 10: §93, §94, §98, §102, §103, §109.
propagines (suckers): 7: §70.
pulpa (pulp): 6: §61, §62; 9: §87, §90; 10: §100.
radiculum (cormlet): 4: §40, §41.
radix (root): 1: §3, §5, §6, §7, §9, §10, §11, §12, §13, §14; 2: §17, §18, §23; 3: §29, §35; 4: §38, §39, §40, §41, §42, §47; 5: §48, §49, §50, §51, §52, §53, §54; 6: §58, §61; 7: §75; 8: §81; 10: §110, §111; 12: §120, §124, §127, §131; 13: §137, §139; 14: §154.
ramus (branch, twig): 1: §11, §14; 2: §17; 3: §35; 4: §42, §45, §46; 5: §49, §50; 8: §82; 10: §110, §111; 11: §116.

– *ramusculus* (little twig): 4: §42; 10: §110.
receptaculum (container): 14: §151.
sedes (implantation): 7: §70, §72; 10: §105, §108, §110; 14: §152, §153.
semen (seed): 1: §5, §6, §9, §11; 3: §28, §29, §37; 4: §38, §42; 5: §51, §52, §53, §54; 6: §55, §56, §57, §58, §59, §60, §61, §62, §63; 7: §64, §65, §67, §70, §72, §74; 8: §77, §78, §79, §80, §81, §82, §83, §84, §85; 9: §86, §87, §88, §90, §91; 10: §93, §94, §95, §96, §97, §98, §99, §100, §101, §102, §106, §108, §110; 12: §120, §122, §124, §125, §127, §133; 14: §145, §146, §148, §149, §151, §152, §153.
siliqua (silique): 5: §51; 7: §64; 10: §97, §99, §101, §104.
singulus (individuality): 4: §41; 9: §90; 10: §96, §97, §105, §106, §108.
spica (clove): 4: §41.
squamas (scale): 7: §72, §74.
stamen (filament [of the stamen], style [of the pistil]): 7: §66, §70, §71, §72.
tunica (tunic): 7: §74, §75; 10: §98.
surculus (scion): 1: §11, §14; 3: §30; 4: §42, §47; 5: §49, §50; 8: §82; 10: §110, §111; 11: §115.
tegmen floris (floral protection, sepals): 7: §75; 10: §96, §97, §98, §99; §103, §104, §105, §106, §107, §108, 14: §151.
vmbilicus (lid, umbilicus): 8: §81; 10: §94.
vasculum (capsule): 7: §64, §70; 9: §90; 10: §97, §98, §99, §100, §102, §103, §104.
virga (rod): 3: §34.